F..., THRIVE, OR DIE

FRY, THRIVE, OR DIE

A Fun Pocket Guide

to 50 Common, Delicious, Hallucinogenic, Medicinal, and Poisonous Mushrooms of the Western United States

DR. MICHAEL AMARANTHUS

gatekeeper press
Columbus, Ohio

The views and opinions expressed in this book are solely those of the author and do not reflect the views or opinions of Gatekeeper Press. Gatekeeper Press is not to be held responsible for and expressly disclaims responsibility of the content herein.

FRY, THRIVE, OR DIE
A Fun Pocket Guide to Fifty Common, Delicious, Hallucinogenic, Medicinal, and Poisonous Mushrooms of the Western United States

Published by Gatekeeper Press
7853 Gunn Hwy, Suite 209
Tampa, FL 33626
www.GatekeeperPress.com

Copyright © 2022 by Dr. Mike Amaranthus
www.frythriveordie.com

All rights reserved. Neither this book, nor any parts within it may be sold or reproduced in any form or by any electronic or mechanical means, including information storage and retrieval systems, without permission in writing from the author. The only exception is by a reviewer, who may quote short excerpts in a review.

Library of Congress Control Number: 2022944239

ISBN (paperback): 9781662931048
eISBN: 9781662931055

These are some gorgeous *Ganoderma conks*, one of the most important medicinal mushrooms, with immune-protective and anti-inflammatory effects. The fruiting bodies are too woody and bitter to eat. People in China, Korea, and Japan commonly drink decoctions of them, a tradition that dates back millennia; I take *Ganoderma* commonly known as reishi in the form of encapsulated extracts and recommend it often to patients.

—**Dr. Andy Weil**
*Clinical professor, Internal Medicine
Director, Program of Integrative Medicine,
University of Arizona*

CONTENTS

Preface	8
Acknowledgments	10
Introduction	11
Using This Book	16
Understanding the Terms	18
Warning!	24
Go for a Walk in the Woods	25
Kids Love Mushrooms	27
Growing Up in a Mushroom Family	30
Sustainable Harvest of Wild Mushrooms	33
Myco-What? Mycorrhizae!	35
What Makes Mushrooms Medicinal?	39
Preparing and Storing Wild Mushrooms	46
Must-Know Mushrooms	49
1. Chanterelles and False Chanterelles	50
2. Milk Caps	72
3. Honey Mushroom	85
4. Amanitas	92
5. Parasols	115
6. American Matsutake	127
7. Prince	137
8. Shaggy Mane	143
9. Psilocybes	150
10. Oyster Mushroom	179
11. Galerina	187
12. Teeth Fungi	190
13. Boletes	208
14. Morels and False Morels	234
15. Truffles	267
16. Corals and Conks	285
Special Dedication to Dr. Jim Trappe, Myconaut	309
References and Websites	315
Index of Common Names	317
Index of Scientific Names	319
Index of Essays, Recipes, History	321
About the Author	325

PREFACE

Wild mushrooms that make a splendid meal; wild mushrooms that can improve your health and maybe save your life; wild mushrooms that can end your life—how do you tell the difference? A friend, Johnny Jones, who has hunted mushrooms for fifty years, tells skeptics who aren't sure how to identify the "fry" (edible mushrooms) from the "die" or (deadly poisonous ones):

"If you can tell the difference between a Pomeranian and a red fox, between a black-and-white cat and a skunk, you can learn to differentiate mushrooms."

We are witnessing a "mushrooming" in interest in fungi. In my experience as a scientist for the last forty years, I have seen so much change. For decades, Americans have had a case of severe fear of fungi ("mycophobia"), germs, parasites, disease, and even the fear to touch a mushroom—let alone eat one. Fungi were the enemy in the management of farms, fields, forestry, parks, and golf courses, where often the only approach to dealing with them was chemical and industrial fungal extermination—basically, nuking the soil and environment into submission. These practices may have delivered some temporary control, but often the harmful fungi would rebel and come back in full force the following year.

Things are changing—it's exciting! Fungal mycorrhizal inoculants are now available for farmers, foresters, and ecologists. Medicinal fungi are available for our health, mind-restoring mushrooms for well-being, and gourmet mushrooms for our palate. Mushroom products are everywhere. Television shows, movies, documentaries, podcasts, and journalists are exploring the fascinating fungal opportunities and functions. Research is exploding into various attributes and uses of our fungal friends.

I hope you will find *Fry, Thrive, or Die* a handy tool in your exploration of the fungal world and mushroom hunting. And welcome aboard . . . because fungi are part of the "all-hands-on-deck" for sustainable planetary solutions!

Reishi products come in a variety of forms.
Research indicates reishi has unique medicinal qualities.

I spent most of my scientific career publishing papers in technical journals that are so full of technical terminology—taxonomic and statistical information—that only other scientists specially trained in the field could understand. Now that I am retired, I love reflecting on the joy of fungal discovery. There is a growing pool of fungi lovers who appreciate what these remarkable organisms can do for our planet. But I suspect you are getting "into" fungi because you want to learn how to hunt wild mushrooms and how to serve them up.

Fry, Thrive, or Die will help you in your quest. With time, you may witness a metamorphosis in your thinking and activities. Your curiosity may be sparked, and you will want to learn more about fungal biology—how they grow and function, how they communicate, how they heal people and the planet, how they stimulate plant growth, and how they recycle the earth's resources. This book is for you.

Enjoy, fry, and thrive.

ACKNOWLEDGMENTS

Everyone "into" mushrooms has had help on their fungal journey. I would like to thank some of those that have helped me on mine.

Dr. Jim Trappe, Dr. Dave Perry, and Dr. Nick Malajczuk—thank you for your passion and insight into the fungal world and your friendships.

To "Myconauts" David Arora and Paul Stamets, thank you for your vision and fearlessness.

I would also like to thank the many who helped make the book fun and insightful with recipes, quotes, stories, comments, anecdotes: Eileen Amaranthus, Dave Steinfeld, Jack Ingvaldson, Eric Ballinger, Dr. Andy Weil, Paul Stamets, Dr. Pam Kryskow, Dr. Megan Frost, Zack Amaranthus, Brianna Amaranthus, Tim Giraudier, Gordy Longhurst, Dr. Jim Trappe, and Chef James Daw. Thanks to Linda Woodrow-Gray for the exceptional illustrations in the book. Thanks to David Steinfeld, Faith Sumalinog, and Aimee Jenkins for the suggestions and insights that make *Fry, Thrive, or Die* understandable and relatable.

A special thank-you to my wife, Eileen, who has excelled in and never doubted this "mycopath" and whose reply to every challenge has been, "Let's do it!"

Go explore your world and find some fungal treasures.

Dr. Mike Amaranthus
Grants Pass, Oregon

Paul Stamets with the author. *Fry, Thrive, or Die* incorporates the insights and passions of many fungal enthusiasts.

INTRODUCTION

Mushrooms! It's perplexing to know what to make of them. Some are great heroes. They can transform an ordinary meal into a great event. They can heal the human body. They can rewire the human mind to see and appreciate the world in new and imaginative ways. But others can be dangerous villains. Some can kill in a matter of hours. How does a person navigate these fungi that, when consumed, can bring forth a munchy, medicinal, murderous, or mind-altering experience? If you are a person wanting to participate in the wonders of the mushroom experience, this book will help you appreciate the differences and personalities of the mushrooms that can make you *Fry, Thrive, or Die*.

The *fungi* word starts with "fun." You will be with interesting people who hunt and gather, party, pontificate, and revel in being outside enjoying nature. The meals will be glorious, the fungal "fish" stories exaggerated, and the memories made . . . unforgettable. That has been my experience.

Fry, Thrive, or Die is full of stories of mushroom hunts, meals and recipes, important medicines, hallucinogens, and ways to store mushrooms, make teas and tinctures, as well as identify and locate must-know mushrooms to make your journey fun and productive.

Interesting people put the "fun" in *fungi*.

Morel discovery puts the "fun" in *fungi*.

For those of you who are new to all this, fungi are neither plant nor animal. They are in a separate kingdom to themselves. It may seem strange, but from an evolutionary standpoint, they are closer to animals on the tree of life than they are to plants.

Dr. David McLaughlin, professor of plant biology at the University of Minnesota, came to this conclusion in the journal *Nature* after analyzing the DNA of animals, fungi, and plants. So if you go to dinner and order a mushroom burger, know that the mushroom is more closely related to the waiter than the lettuce on the bun. Crazy.

Fruiting bodies of fungi in this book are referred to as mushrooms, corals, conks, or truffles. They are the "apple," and the fungal threads, or *hyphae*, in the soil or wood are the "apple tree." These hyphae are efficient enzyme producers, and like the microbes in human stomachs, they digest food in the soil, root, or wood for nourishment.

Hawksworth and Lücking (2017) in *Fungal Diversity Revisited* estimate that there are 2.2 to 3.8 million species of fungi. Regrettably, it's estimated that only fourteen thousand produce mushrooms (Miles and Chang 2004). But all these fungal species are still busy doing something. Some are critically important for making bread, cheese, wine, beer, and medicine; sequestering carbon; recycling nutrients; and caring for plants. Some are known pathogens. For the millions of other fungal species, we are just beginning to learn about them and what they do. I have often wondered why most major universities have departments of plant science but not departments of fungal science. There is so much left to learn about fungi.

You will need to know your mushrooms—not all of them, but the ones you want to use. It's not rocket science, but it does take some experience and time. Hang out with people that have mushroom-collecting experience. It seems every community has mushroom clubs and classes to help you on your journey. And if your community doesn't, start one yourself.

Life Learning

Using this book should increase your knowledge and appreciation of these over fifty common and important mushroom species. Obviously, including all Western US mushrooms in a pocket guide would require pockets three feet deep. But if you are curious about the other fungi you may encounter on your forays, there are technical guides that cover the hundreds of species not described in this book. Two comprehensive and technical references are *Mushrooms Demystified* by David Arora and the *National Audubon Society Field Guide to North American Mushrooms* by Gary Lincoff. I have included a list of handy references and websites at the back of this book for the mushroom collector and enthusiast wanting detailed information on a variety of fungi and their uses. The information on fungi continues to evolve and expand. Enjoy the journey.

Mushroom People

Mushroom people are some of the most unique people to be found on the face of the earth. Different in many ways from the people around them, they speak a strange language of spore prints, caps, rings, gills, bruising reactions, and mysterious fungal forms. They wander and ponder endlessly about subtle differences in habitats and microclimates. Curious, passionate, and wondrous adventurers, mushroom people are a crazy cross between a gold rush seeker and Alice in Wonderland. They can be found on every continent and from every background with a twinkle in their eye and a bag on their hip. They are the mushroom people.

USING THIS BOOK

Fry, Thrive, or Die is a pocket field guide to distinctive and important wild mushrooms found in the Western United States. The over fifty mushrooms in *Fry, Thrive, or Die* were selected because they are delicious edibles, medicinally important, or dangerous. These fungi extend from the Pacific Coast to the Rocky Mountains and from Mexico to the Canadian border and beyond.

You can use the keys in the back of the book for gilled mushrooms and for non-gilled mushrooms to determine if you have picked one of the fifty-plus mushrooms. Once you collect a mushroom that you suspect is one of these mushrooms, use the key to narrow down the species and location described in the book. At this point, it is important to compare the characteristics of your mushroom with the description and illustration of the species you suspect it is. To make a positive identification, *all* the key features of your mushroom must be present as described for that species, and if they are not, you must assume it is not the correct mushroom. It's best to gather several specimens so you can look at the range of characteristics as the mushroom ages. For example, the color of the gills is a common characteristic that can change as a mushroom matures.

At the top of each mushroom description is an icon—these designating whether it is a "fry," "fly," "thrive," or "die" mushroom. The icons stand for the following:

- **Fry:** Edible[1]
- **Fly:** Hallucinogenic
- **Thrive:** Medicinal
- **Die:** Can cause severe gastric stress and in some cases kill you

---

[1] WARNING: Some people get allergic reactions or stomach distress from eating mushrooms that are commonly consumed as edibles. So go easy eating a "fry"-designated mushroom for the first time.

fry — **thrive** — **die** — **fly**

Several mushroom species have more than one icon. For example, some mushrooms are known to be good edibles (such as hen of the woods and lion's mane) and are also used medicinally, hence both the "fry" and "thrive" icons. Some mushrooms that are hallucinogenic (such as the fly agaric and panther amanita) can also create severe gastric distress and nausea for some people depending on dosage and sensitivity. These mushrooms are given both hallucinogenic ("fly") and poisonous ("die") designations, even though eating a fly agaric or panther amanita can you make you very sick but will not kill you.

UNDERSTANDING THE TERMS

It can take some experience to fully understand mushroom descriptions. For some mushrooms, the presence or absence of a universal veil or a ring on the stalk is critically important. For others, the key characteristic could be whether the gills are attached or free from the stem, and still for some, the presence or absence of a volva. For yet others, the color of the gills or smell can be diagnostic. I've included some important pinpointing features in the description heading for each listed mushroom and illustrations that highlight features that will help you identify a specific mushroom species.

The warts on an *Amanita muscaria* cap.

Underneath a Pacific golden chanterelle cap are blunt interconnected folds and not sharp-edged blades or gills.

Common and Scientific Names

Each mushroom description begins with its common name followed by the Latin scientific name for the genus and species. Other known common names are also listed.

Cap Characteristics, Underside of Cap, and Stalk

These are the features most used to distinguish and identify the mushroom.

The Veil and Ring

The veil is a tissue layer that extends from the edge of the young cap to the stalk. As the cap expands with age, the veil breaks, forming a ring on the stalk. Some mushrooms form a "universal" veil that completely encloses the young mushroom like an egg. Once this veil is ruptured, it creates a saclike "volva" at the base of the stalk, leaving warts and patches on the cap. *Amanita* species have a volva, and it's important to dig under the base of the mushroom stalk to determine if the volva is present.

Size

A large cap is six inches or greater. A medium cap is two to six inches across. A small cap is less than two inches. A thick stalk is greater than one inch in thickness; medium thick stalk, 3/8 to one inch in thickness; and thin stalk, less than 3/8 inch in thickness. The size of a mushroom is often a function of its age.

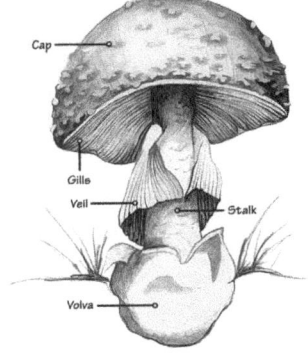

A. Free
(gills not attached to stalk)

B. Adnexed or Adnate
(gills narrowly or broadly attached to stalk)

C. Decurrent
(gills running down the stalk)

Gills and Gill Attachment

Some gills are widely spaced, and some are tightly spaced. Some gills are deep and knife-edged, and some are shallow, thick, and blunt-edged (like folds, e.g., chanterelles). Gills attach to the stalk in many ways. Free gills are not attached to the stalk. Adnexed gills are narrowly attached to the stalk. Adnate gills are broadly attached to the stalk. Decurrent gills run down the stalk.

Gill Color and Spore Print

It's important to note the gill and spore colors when identifying mushrooms. Spore color is determined by taking a spore print. It's easy. This is how you do it:

Take a mature mushroom, and place it on a piece of white paper (dark paper if the gills are white) with the gills facing down. Cover it with a glass or bowl for several hours. The spores will fall from the gills, leaving a distinctive pattern. Boletes and hedgehogs will also produce spore prints. The spore color can also be determined by looking at the spore dusting on the veil, stalk, or cap of the mushroom being identified. The gill colors can change with maturity, so get a range of mushroom ages with your collections when possible.

Spore prints are important diagnostic features of mushrooms.

Fragrance and Edibility

Mushrooms contain a wide diversity of smells and tastes that vary widely in type and intensity. For most edible mushrooms, I've included some favorite recipes in this book and some excellent cookbooks to consider in the References and Websites section.

Many mushrooms, like the morels, are edible and delectable.[2]

---

[2] Some individuals have allergic reactions or stomach distress eating mushrooms that are commonly consumed as edibles. Best to eat a small portion when consuming a "fry"-designated mushroom for the first time. More people get sick from overindulgence or eating rotten mushrooms than actual mushroom toxins.

Habitat

Where a mushroom is found, when it fruits, and how abundant a crop may be is covered in the "Habitat" sections. There are many factors that go into a mushroom's fruiting habitat, and with experience, you will begin to understand some of these factors as they pertain to locating your favorite mushrooms in time and space. You may learn to associate certain types of vegetation or disturbances with the mushrooms of your choice in your area.

In the Western United States, forest and mushroom habitats are often as far as the eyes can see.

Edible mushrooms and mushroom hunters are often found in clustered patches and not evenly distributed across the landscape.

Similar Mushrooms That Should Be Avoided

I've included common mushrooms that look similar and could be confused with the fifty-plus mushrooms described in this book. Obviously, these look-alikes should not be consumed because they can make you sick or even kill you. Depending upon a collector's experience and knowledge, there could be a much larger group of "look-alike" mushrooms. In addition, some people have allergic and gastric reactions to mushrooms that other people consume safely. Bottom line is to be safe and certain of the identification of any mushroom you consume. *When in doubt, throw it out!*

WARNING!

This book is intended to encourage the reader to go out and discover mushrooms. There are over a thousand mushrooms in the Western United States. This book covers over fifty. This book is not a "know-it-all" guide to harvesting, eating, cooking, and processing wild mushrooms. Nor is this book intended to guide the reader into consuming mushrooms for culinary, medicinal, or mind-altering experiences.

It is important to remember that some mushrooms are poisonous and a few are deadly. I've seen deadly mushrooms grow within patches of delicious mushrooms, so examine all specimens. Proper identification is essential. If you are an amateur, find experienced experts to learn from. I will provide some excellent books and online references to learn more details regarding the identification and uses of culinary, medicinal, or mind-altering mushrooms. Many mushroom poisonings are caused by people eating spoiled mushrooms. Mushrooms keep more like fish and less like vegetables. When in doubt, throw it out. And please, until you get experienced, never eat a white-gilled mushroom or little brown mushrooms. Some deadly *Amanitas* have white gills, a veil, and a volva here in the Western United States. They could put you in the "die" category in the book's title *Fry, Thrive, or Die*.

This book will help educate you regarding some important mushroom species in the Western United States. It does not replace the advice of a physician or qualified health professional. Some people have unique food allergies and sensitivities. The author takes no responsibility and does not encourage anyone to ingest any mushroom solely based on the information in *Fry, Thrive, or Die*. The author and the publisher are not responsible for any adverse effects or consequences from the use of information in this book.

Some mushrooms, such as the *Amanita ocreata*, are deadly poisonous.

GO FOR A WALK IN THE WOODS

What always makes my day better? Going for a walk in the forest and looking for mushrooms! Sometimes I find just a few, sometimes I find the mother lode, and sometimes I get "skunked." But I always enjoy the natural beauty in my surroundings.

In the woods with a mountain blond morel.

Walking in the woods is a great activity to do with family and friends. Finding mushrooms is like a treasure hunt to be shared with people. Mushrooms vary widely in color, shape, size, and smell. Some mushrooms grow on litter and dead leaves, some on living roots near the soil surface, and some on rotting wood and standing trees. So it's never boring and always challenging to determine what you have discovered. This book will help.

Furthermore, people aren't the only ones who like finding mushrooms. The woods are full of *mycophagists* (critters that eat mushrooms). Because mushrooms contain protein, minerals, vitamins, and other beneficial compounds, many wildlife species search out and ingest mushrooms. Squirrels, voles, deer, elk, cougars, and bears are just a few of the mycophagists that consume mushrooms. I know this because I have run into them in my forty-plus years of wandering in the forests, and I have seen the remains of their feeding—teeth marks in porcini, scattered remains of orange chanterelles, and rotten logs and duff torn apart by bears and squirrels looking for truffle fungi. And what do the mushrooms get in return for being eaten? Mushrooms use wildlife to spread their spores in fecal matter that gets deposited across the forest floor.

Okay, by now you know I'm crazy about mushrooms. Millions of people are catching the mushroom "bug." They are incredible organisms. Fungi produce bread, beer, wine, cheese, and medicines that make our world a better place. Some mushrooms glow in the dark, some transform our brains, some are our most expensive meals, and some are the oldest and largest living things on the planet. This book is a celebration of their diverse and remarkable contributions.

So go out and explore. It all starts with a walk in the woods.

KIDS LOVE MUSHROOMS

We are all descendants of hunters and gatherers, and our bodies and minds have evolved for that purpose. Children, over millennia and once they are old enough to forage, were important parts of the survival of family units. Today, the first time kids find a valuable plant or mushroom is likely a time of celebration. Kids love being producers. It gives them a sense of purpose, importance, and excitement. Watch the look of joy on a child's face when they find a morel, porcini, or a reishi, knowing it will be an honored member of the family table or the medicine cabinet. There is a sense of happiness when children learn about their world and discover that their actions have meaning.

It's interesting how different generations feel about mushrooms. When I was a kid, the Amaranthus family would often assemble on Sundays at my Italy-born grandpa's house for dinner. He would make a big meal for his five children, their spouses, and fifteen or so grandchildren, including me. The main dish would invariably have wild mushrooms included. I vividly remember my mother and aunts, who were born and raised in America, painstakingly pulling the mushrooms off the plates of the grandchildren. They feared my grandpa's wild mushrooms would poison us.

I think the fear of mushrooms is primarily projected by the attitudes of adults. My own five kids and eight grandkids have grown up around mushrooms. In our house, there are always some fresh edible, medicinal, or "curious" specimens in the fridge. Medicinal tinctures and teas are in a pot on the stovetop after a morning brew. Dried and frozen fungi are stacked in the pantry and freezer. I have a microscope in my home office, and the kids and grandkids look at specimens and spores when they come to visit. So basically, there is little fear of mushrooms in our family. We raised our kids to hunt mushrooms, and we still go out with the kids and grandkids in the spring and fall. Not all kids love to eat mushrooms, but they all love to get out into the forests and fields with their bags and field gear. They get dirty and find treasures. They see butterflies, wildflowers, and wildlife, and they hear stories of the mushroom hunting past. There is something about hunting and gathering that resonates deeply into the human soul.

GROWING UP IN A MUSHROOM FAMILY

Zack Amaranthus

Anyone who has ever walked through a home improvement store at some point in their life has surely seen an inconspicuous tool called a potato hoe; it is used to pop potatoes out of the ground. It's one of those tools that will leave you guessing, "What the hell would anyone use that for?" It's got a long handle and a short head with curved, four-inch tines. Oddly enough, my family always had seven potato hoes in the garage, yet we never grew potatoes . . . ever.

Geographically speaking, our family lived on the side of a mountain surrounded by a forest with a view as far as the eye could see. Not exactly a great place to grow potatoes. My dad being a scientist and me being a curious kid led me to ask the question, "How do these trees grow so tall without anybody watering or fertilizing them?" And to his delight, he answered with just a single word, "Mycorrhizae!" A typical parent would probably have answered that question by talking about the annual precipitation or our geographical location in the Pacific Northwest. Not my dad. He knew there was something going on underneath our feet that was a keystone to the growth of trees. Without even knowing it, my childhood interest had been written with that single word. The relationship between fungi and plants would strangely play a part in my career choice later in life.

My first mycorrhizal exploration came at an early age. The potato hoe, a tool my family dubbed a "truffle fork," finally had a purpose.

To this day, if you ask any one of my siblings, they wouldn't have a clue what a potato hoe was even though they've all handled one a hundred times. A truffle fork, on the other hand, everyone knows that this is a crucial tool to our family. It is a fork, essential to pulling back just enough soil to find the most prized mycorrhizal fruiting body, the elusive truffle.

My parents bought me a toy "truffle fork" for my second birthday.

Zack at two years old holding a truffle.

Our discoveries were not the white or black truffle of culinary fame but a *Rhizopogon* truffle. Our family called them "pogies." My folks used these fungi to inoculate trees in their forestry inoculation business to improve the growth and survival of tree seedlings. Hunting them was never a crapshoot. We were trained to only look where there was a squirrel pit: a small hole in the forest duff where an animal had dug. When we found one, we would use our forks to examine an area about three inches deep and four inches wide in either direction in search of the small, potato-like truffles. We were careful to replace the forest floor. These trips were meticulously mapped by my dad. He knew which tree species formed truffles and the time of year they would form. When we found truffles, my mom and dad would make a big deal over our treasured discovery, and there was usually ice cream to celebrate a successful hunt.

Seasonally, my siblings and I became hunters and gatherers. It was fun to be out in the woods competing in these "hidden easter egg hunts." It's interesting that thirty-five years later, my life revolves around fungi, from producing mycorrhizal products to growing truffle-colonized trees. My life is now dedicated to fungi. I eventually stopped questioning the number of truffle forks in the garage. Those forks lined the wall of a mushroom-loving family.

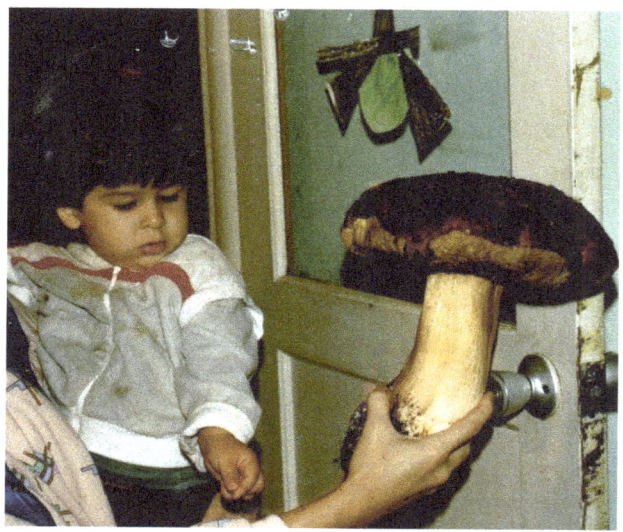

Zack aged two with a giant porcini.

SUSTAINABLE HARVEST OF WILD MUSHROOMS

Throughout history, mushrooms have been important sources of food and medicine. In the Western United States, recreational and commercial harvests have grown dramatically in recent decades. Supplemental income of thousands of individuals working full time or part time in the commercial harvest of chanterelles, boletes, morels, and matsutake has contributed millions to local economies (Amaranthus and Pilz 1996).

Today thousands of recreational and commercial pickers continue to harvest wild mushrooms from public and private lands. Yet there is uncertainty about the ecology of wild edible mushrooms, the effects of climate change, and the intensity of disturbances caused by wildfire, drought, and disease (Perry and Amaranthus 1990). At the center of the management issue is a lack of information on productivity and habitat requirements, interaction of forest health, and the effects of repeated mushroom harvest (Amaranthus 1998).

Some encouraging data exists regarding managing forest habitats to increase mushroom production is possible for some species (Amaranthus et al. 1998). In addition, other data suggest that careful mushroom harvest without excessive disturbance of the forest floor does not negatively affect wild mushroom production (Amaranthus et al. 2000). Because mushrooms and truffles are just the fruits of the mycelium, they can—if done properly—be harvested in a sustainable way by only taking the fruit and leaving the mycelium intact.

More data is needed as wild mushroom harvest continues to expand. Clearly, if wild edible and medicinal mushroom harvest is to be sustained, it must be done by understanding the biological, economic, and social forces behind their harvest. So when you are out collecting your wild mushrooms, pick only the healthy specimens (leave old or rotten fungi in the woods). Also, use a mesh bag. The holes in the mesh allow fungal spores to fall to the ground and spread across the landscape with you. Don't excessively disturb or rake the forest floor and the fungal mycelium in search of young specimens that are hidden from view beneath the duff layer. Remember also to replace the duff layer on the forest floor after it has been removed.

This matsutake mushroom was carefully excavated and harvested with minimum disturbance to the forest floor and mushroom mycelium.

MYCO-WHAT? MYCORRHIZAE!

Mycorrhizal mushrooms connected to trees.

Many of the edible mushrooms described in this guide are mycorrhizal. They include chanterelles, boletes, milk caps, matsutake, hedgehogs, black trumpets, truffles, and many morel species. And some of the most poisonous mushrooms—like *Amanita ocreata*, *Amanita phalloides*, and *Gyromitra esculenta*—are mycorrhizal too.

Technical words, like *mycorrhiza*, get thrown around a lot by scientists. Naming things, in scientific ways, is the basis for beginning to understand how things work. But at this point, the real question you may be asking is, "Myco-what?" Scientific terminology and Latin names can be confusing as we look for fungi. None of this must be so technical, so full of jargon and gobbledygook that we cringe at the sound of certain terms like *mycorrhiza*. I am going to suggest, however, that *mycorrhiza* is one term you learn so you can gain a greater appreciation for the mysteries (and the location!) of the mushrooms you are picking.

Mycorrhiza is probably the best-studied plant-microbial relationship. There are over a hundred thousand peer-reviewed scientific studies in the technical literature. The problem is the scientific literature is difficult to access and understand, especially without the experience of reading technical papers. It is time to explain the basic workings and benefits of this remarkable group of fungi in plain terms.

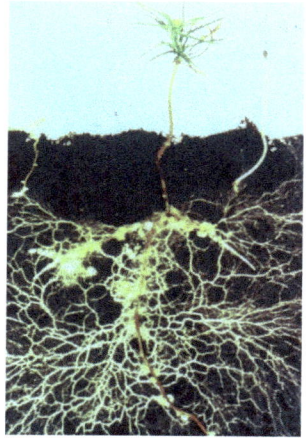

White mycorrhizal filaments extend from a pine tree grown in a glass box.

Myco-What?

Let's start with some basics. Fungi form a beneficial living relationship with a vast majority of plant species. This is a symbiotic relationship where both species work together to survive and thrive. It's a win-win. We call a root-fungus combination a *mycorrhiza* and its plural, *mycorrhizae*. *Mycor-rhiza* literally translates to "fungus-root." The roots of an estimated 85 percent of the world's plant species have this association with these specialized fungi (Rimington et al. 2020).

Mycorrhizal fungi form the network of fungal filaments, or "threads," that permeate into the soil from their home on the plant root. The body of the mycorrhizal fungus consists of microscopic filaments called *hyphae*. An individual *hypha* (singular) is approximately 1/25th the diameter of a human hair and can grow up to two feet in length! Hyphal strands grow from within and around the root cells of the "host" plant, spreading out into the surrounding soil. These same hyphae can aggregate into a collection of hyphae, called *mycelium*, which forms many of the mushrooms and truffles described in the *Fry, Thrive, or Die* field guide.

The trading of soil water and nutrients, captured by the mycorrhizal fungus for sugars produced by plant photosynthesis, is the foundation for this ancient relationship dating back 460 million years. Mycorrhizae in nature are the rule, not the exception, and they are fundamental to plant nutrition for most plant species. Putting this together in the big picture, this is how the living soil got started and the reason plant life began to flourish across a once-barren earth!

Mycorrhizal fungi are particularly important in accessing phosphorus, nitrogen, zinc, iron, calcium, magnesium, manganese, sulfur, and other important soil nutrients (Clark and Zeto 2008). They also help protect the plant root system against drought and disease and help plants revegetate degraded sites (Amaranthus et al. 1989).

The trees on the right were colonized by mycorrhizal fungi and have better nutrition.

Mycorrhizal Fungi Form Mushroom- and Truffle-Fruiting Bodies

For the mushroom hunter, however, you are most interested in mycorrhizal species that form mushrooms. The species are associated with conifers, oaks, pecans, hazelnuts, beeches, eucalyptus, alders, cottonwoods, poplars, birches, and some tropical hardwoods. When you see a mushroom, you can think of it as the "fruit" of a much larger fungus, much like the apple is to the apple tree.

A mycorrhizal mushroom fruit from a container where the tree is supporting its fungal mycelium.

Well over four thousand species of mycorrhizal fungi occur in forests across the globe. When you walk through the woods and find a mushroom or a truffle, you are probably seeing the "fruiting body" of an ectomycorrhizal fungus.

Aboveground, a mushroom disperses its spores into the air, while belowground, the hidden truffle uses another strategy: it attracts mammals or other animals with a hormonal scent to move its spores around. Once consumed and dispersed, these spores can colonize the sites of new tree roots. Truffle species are also known for their epicurean delicacy, yet not all truffle species are tasty. The nonedible truffles are quite common in Western Oregon and Washington, and their production can actually exceed that of mushrooms on some sites (Amaranthus et al. 2001). Tasty or not, they contain millions to billions of spores, or "fungal seeds," waiting for a tree root to colonize.

WHAT MAKES MUSHROOMS MEDICINAL?

Mushrooms have been used medicinally for thousands of years, but it is only recently that scientists have documented some of the specific mechanisms for how mushrooms improve health and vigor. Several well-documented medicinal mushrooms (such as reishi, hen of the woods, lion's mane, oyster mushroom, chaga, and turkey tail) are featured in *Fry, Thrive, or Die*.

Medicinal mushrooms contain certain polysaccharides, polyphenols, and antioxidants that are scarce in other foods. Many have high concentrations of vitamins, amino acids, nutrients, and micronutrients. These beneficial characteristics can help the body reduce inflammation and cholesterol, as well as combat tumors, viruses, and bacteria. If you want to delve into the current research on mushrooms, check out the mushroomreferences.com website. It contains hundreds of abstracts and citations of peer-reviewed scientific studies related to the medicinal qualities of specific mushrooms. The website lists studies by mushrooms species, and it is an effective way to keep up with the latest scientific research for chaga, lion's mane, hen of the woods, *Psilocybe*, reishi, turkey tail, and many more.

Some mushrooms, like the hen of the woods, are medicinally important and also delicious to eat.

In this section, I will cover some of the important compounds found in medicinal mushrooms and what they can do for human health. Then I will guide you on how to prepare them in teas and tinctures. It is important to know how these compounds are released from mushrooms. Some of these compounds are soluble and released in hot water (for example, in preparation of teas) while others are only soluble in alcohol (for example, when making tinctures). Still others are released partially in water but more thoroughly extracted with alcohol, and in these cases, tinctures can be made that extract both water-soluble and alcohol-soluble compounds. There are also some details on the specific compounds, but if you are not interested in such details, you can move on to the "Where Do I Find Medicinal Mushrooms?" section.

Polysaccharides

Mushrooms are known for containing long-chain carbohydrate molecules called *polysaccharides*. Certain polysaccharides (such as beta-glucans, chitin, and arabinoxylan) are known to boost the immune system, inhibit tumor growth, and slow inflammation. Beta-glucans are perceived by the body as foreign invaders. As a result, the body produces an array of defense mechanisms, including production of cytokines, macrophages, and natural killer cells (T cells), which make the immune system more responsive to pathogens. Arabinoxylan is also present in many medicinal mushrooms. It is a major component of dietary fiber, which improves digestion and protects the body from infections. Polysaccharides are water-soluble.

Melanin

Melanin is a pigment produced by most living organisms. For many fungi, it is produced in abundance. Melanin acts as an armor, protecting the fungi from ultraviolet radiation. It works similarly in humans where melanin is present in the hair, eyes, and skin. Melanin from fungal sources can improve the health of the epidermis layer of the skin and help maintain the healthy pigmentation of the skin and hair. It also forms a melanin-glucan complex that scientists have identified as providing high antiviral protection against a variety of viruses. Melanin is water-soluble.

Vitamins B and D, Minerals, and Amino Acids

Mushrooms are rich in B vitamins: riboflavin (B2), folate (B9), thiamine (B1), pantothenic acid (B5), and niacin (B3). These help the body utilize energy from the food we consume and produce red blood cells, which carry oxygen throughout the body. Medicinal mushrooms are a great source of vitamin D. When mushrooms are exposed to sunlight, they convert *ergosterol* (a sterol produced in mushrooms) into vitamin D. Medicinal mushrooms contain nutrients and minerals (such as phosphorus, copper, potassium, magnesium, zinc, and selenium) that are essential for good nutrition. A variety of medicinal mushrooms also contain essential amino acids (leucine, lysine, histidine, methionine, phenylalanine, threonine, tryptophan, and valine). Vitamins, minerals, and amino acids are water-soluble.

Polyphenols

Polyphenols are a large group of organic compounds prominent in fungi. Polyphenols are considered antioxidants and protect cells from the oxidative damage of free radicals. The antioxidant value of certain mushrooms can be very high. For example, the antioxidant value of chaga is 1,500 times higher than blueberries or blackberries. A long-term study from Spain published in 2021 found that certain foods rich in polyphenols (which include mushrooms as well as coffee, cocoa, and red wine) may be protective against cognitive decline in older adults. Polyphenols can be extracted by both water and alcohol.

Ergothioneine and Glutathione

Mushrooms contain a very high concentration of ergothioneine and glutathione. When these antioxidants are present together, they work to protect the body from the physiological stress that causes aging. Penn State researchers found that the antioxidants ergothioneine and glutathione may also help prevent Parkinson's and Alzheimer's. Ergothioneine and glutathione can be extracted by both water and alcohol.

Triterpenoids

Triterpenoids are a class of chemicals officially classified as being composed of three terpene units (*terpenes* are aromatic compounds). Many medicinal mushrooms have high concentrations of triterpenoids. Triterpenoids are used for medicinal purposes in many Asian countries for anti-inflammatory, analgesic, antiviral, and antitumoral applications. Triterpenoids are a complex of many beneficial compounds, such as sterols, betulin, and ergosterol. Triterpenoids are best extracted with alcohol.

Where Do I Find Medicinal Mushrooms?

You will be surprised where you might find medicinal mushrooms once you start looking. I have found them in neighborhoods, golf courses, parks, and forests. Ironically, I have spied them in front of doctor's offices, hospitals, and Safeway parking lots too. The city, the suburbs, and the "boonies" all have potential. What you need are trees, downed wood, stumps, or buried logs and branches. Some medicinal mushrooms prefer mature forests; some like oak scrublands; while others are just over the fence in your neighbor's backyard. When you are bored or waiting for someone, go outside and look for them. I guarantee it will be more rewarding than checking your smartphone. The descriptions of species in *Fry, Thrive, or Die* will help you in your quest for medicinal offerings found in a variety of habitats.

Chaga fruiting on a birch tree in a neighbor's yard.

Mushrooms at a growers' market in Oregon that provides fresh and dried medicinal and edible mushrooms year-round.

Artist's conk reishi growing in a mature forest.

Making Teas and Tinctures

Once you have collected your own medicinal mushrooms, and discarded any spoiled or moldy ones, you can begin to prepare your specimens for making teas and tinctures. Large, woody conks—like reishi and chaga—need to be processed into smaller chunks. For fresh specimens, you can split them with an axe or hatchet. For dry ones, simply place them in large ziplock bags that are double-bagged and hit them with a hammer. Turkey tail mushrooms are a little easier and can be cut up with shears or scissors. Your neighbors will likely be wondering what you are doing whacking away on mushrooms in your yard. Sure, they might think you are crazy, but really, do you care at this point?

If you can't find medicinal mushrooms in the wild, you can always buy them at health food stores or online. They come in bulk powder, capsules, and tinctures. Many growers' markets and grocery stores now carry healthy, fresh, and delicious shiitake and oyster mushrooms. In addition, hen of the woods and lion's mane are wonderful culinary delights and are getting easier to find in gourmet and supermarkets as well. As Hippocrates, the famous Greek physician, said in 440 BC, "Let food be thy medicine and let thy medicine be food."

Fresh or dried medicinal mushrooms can be used as teas or tinctures.

Preparing tinctures with medicinal mushrooms chaga and reishi.

Teas

A simmer or boil is necessary to unlock the polysaccharides, melanin, vitamins, minerals, amino acids, and some polyphenols compounds found in mushrooms. Boiling the mushrooms for ten minutes releases these compounds and breaks down chitin, which is present in mushroom cell walls. Chitin is the same material that composes crab shells because, as noted before, mushrooms are more closely related to crustaceans and insects than plants. Use one-part mushrooms to twenty-parts water. If you make a big pot, put it in the refrigerator, and the tea will keep for several days.

Tinctures

Tinctures are generally an extraction of both water- and alcohol-soluble compounds. Pack a glass jar (don't use plastic) with your medicinal mushrooms, and cover with high-percentage alcohol, like Everclear or vodka. Seal the container, and shake it once a day for two to four weeks. When you are ready, decant the infused alcohol through a filter cloth, and squeeze the remaining mushrooms to extract all the liquid. Take the squeezed mushrooms and cover with water. Boil for fifteen to twenty minutes. This liquid water extract can then be added to the alcohol infusion at a 5:1 ratio. Tinctures are stable and can be kept for at least a year at room temperature. You can add two full droppers of tincture to water, juice, coffee, or any other beverage morning and night.

PREPARING AND STORING WILD MUSHROOMS

Mushrooms are fungal "fruits," and they should be stored and processed like berries or other fruits. When gathering mushrooms, pick only the healthy-looking specimens. These are the young or recently mature mushrooms, like the ones you might select when you go shopping for them in the supermarket. Mushrooms that are buggy, gooey, overly mature, or have a foul smell should be left in the field. That way they are a food source for other animals, and their spores will disperse to seed future mushrooms. Fresh edible mushrooms have a rich earthy smell, each with their own tones and fragrances. Like berries, do not wash mushrooms until you are ready to prepare them. This is a common mistake and a big no-no! You don't want to turn your wild mushrooms into "mush."

To prepare, clean each mushroom. This is a time where you can admire your harvest. If there are things attached to your specimens (like soil or needles, twigs, or grass blades), clean them off with a soft brush. Then store the mushrooms in a paper bag in the fridge. Storing them in a plastic or sealed container will quickly make them slimy. Fresh, good-quality wild mushrooms can be stored in a paper bag in the fridge for several days. But remember, your fungal fruits are more like fish than vegetables in terms of length of storage. If your mushrooms do start smelling foul or "fishy," dispose of them. More people get sick from eating spoiled mushrooms than from eating poisonous ones.

Boletes and morels can be dried and stored for later use. They rehydrate well and taste as good (some say better) as the original fresh mushrooms. For long-term storage, morels and boletes should be sliced (morels in half longways and boletes in quarter-inch-thick slabs) and placed on a dehydrator. In a few hours, they are ready for long-term storage in a sealed container. Boletes will take longer to dry than morels.

Placing morels on a dehydrator tray.

For mature boletes, you can separate the pores (peel the layer of tubes from underneath the cap) easily from the cap and dry them separately. When rehydrated, the dried edible bolete pores are fantastic in sauces and soups. Both morels and boletes rehydrate nicely when placed in warm water for thirty minutes, and the lovely flavors of these fungi will intensify. Some people don't have dehydrators, so they dry their morels by stringing them together with a needle, flossing through the stem, and hanging them, like prayer flags, outside on warm days. You can also chop them fresh and place them in ziplock containers in the freezer. When needed, they are simply thawed and sautéed. Like all mushrooms, they should be cooked before eating.

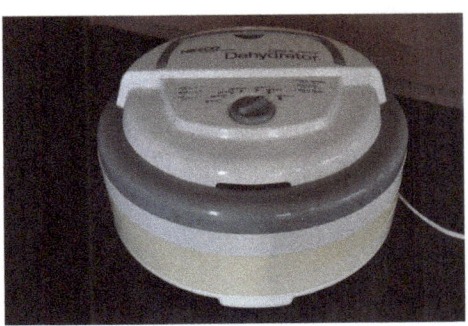

A dehydrator.

For chanterelles, drying makes them too tough and chewy, so preparation and storage methods are different. To store chanterelles, they should be thoroughly cleaned with water and broken into one-eighth- to quarter-inch-thick strips for immediate dry-sautéing. Place them in a hot pan without oil or butter for approximately five minutes to drive excess water from the mushrooms. Place the sautéed chanterelles on a towel for air-drying and cooling. Then put them in a sealed container and freeze for later use. You can even use the chanterelle liquid that remains in the pan. You can use it in sauces or freeze it as stock for your yummy mushroom soup.

When cooking fresh wild mushrooms, wash the mushrooms right before you cook them. Mushrooms can be sautéed, roasted, broiled, grilled, barbecued, or boiled. Mushrooms should not be eaten raw. Bon appétit!

Dried morels rehydrate to almost their original size, gracing any meal year-round.

MUST-KNOW MUSHROOMS

1. CHANTERELLES AND FALSE CHANTERELLES

Popular, widespread, and abundant, a few good chanterelle patches can fill your refrigerator and freezer with collections. They tend to fruit every year in the same exact spot, so good recon can pay dividends for decades! They taste fruity, and the Pacific golden and white chanterelles are large and dense. There is one close look-alike for the collector with some experience: the false chanterelle (also featured in this book for handy comparison). Get to know it, and eat comfortably and joyfully with friends and family.

This group is distinguished by its trumpet-shaped or wavy, vase-shaped cap and blunt folds rather than thin-bladed gills beneath the cap. The mushrooms in this section can be distinguished from each other by their color and the underside of their cap, which generally have blunted interconnected veins (or, in the case of the black trumpet, is smooth). The false chanterelle is distinguished by the prominent scales on its cap and the contrasting color between the reddish-orange cap and the white to pale-yellow stalk. The yellow foot chanterelle is small, delicate, and has a hollow stalk. Chanterelles and the false chanterelle are mycorrhizal with trees.

Finding chanterelles are like finding pots of gold in the forest.

There are ninety species in the broad chanterelle group worldwide with about forty species in North America. Chanterelles are popular edible wild mushrooms of considerable economic value. I have highlighted four delicious widespread and abundant species: the Pacific golden chanterelle, the white chanterelle, the yellow foot chanterelle, and the black trumpet. The false chanterelle is included because it resembles the Pacific golden chanterelle and occurs in similar habitat but causes gastric distress if consumed.

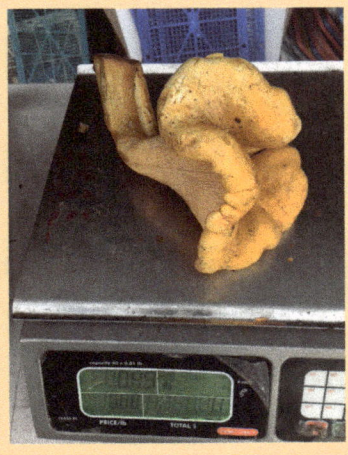

Chanterelles are popular edibles all over the world and important economically.

PACIFIC GOLDEN CHANTERELLE

(Cantharellus formosus)

The Pacific golden chanterelle (*Cantharellus formosus*) is a very common and highly sought-after edible mushroom found in the Pacific Northwest and Northern California. It is a very large, meaty mushroom in the Cascade Mountain Range and in California, but a smaller and more delicate variety occurs in the coast range of the Pacific Northwest and in the Rocky Mountains. The Pacific golden chanterelle is an important commercial species because of its economic value and abundance. It is tasty in soups or as a sauce over meat, poultry, or pasta. It is popular in specialty markets and restaurants, and you can often find them in the produce section of your local grocery store. But can you afford to buy them? They are indeed spendy when purchased fresh in retail stores. If you go find your own patches of chanterelles in the fall, you can return to the woods year after year in the same place to collect your own golden treasures.

The Pacific golden chanterelle goes by other names—including "golden chanterelle," "yellow chanterelle," "girolle," "pfifferling," and *Cantharellus cibarius*.

Cap Characteristics

The cap ranges from small to large (one and a half to seven inches across) and thick. The cap is smooth, not scaly, and colored orange to yellow. The top of the cap is sunken or depressed, funnel-shaped, and usually wavy.

Underside of Cap

The cap underside is orange, yellow, to slightly salmon-colored with blunt veins. The veins are widely spaced, sometimes interconnected, and generally lighter in color than the cap. The veins also extend down the stalk (decurrent).

Stalk

The stalk is generally one to four inches high and the same color or slightly lighter than the cap. Blunt veins, or wrinkles, run down the stalk, often interconnecting. The stalk slightly tapers down from the cap and is solid, not hollow.

Fragrance and Edibility

Pacific golden chanterelles are fragrant and fruity, with hints of apricot and pumpkin. It's an excellent edible mushroom that can be often found in abundance.

Habitat

You can find Pacific golden chanterelles in conifer forests of Oregon and Washington and Northern California. It fruits from July to December. In the Pacific Northwest, it had been classified as *Cantharellus cibarius* (the golden chanterelle from Europe); however, it was determined through DNA testing to be a distinct species, and it has recently been reclassified as *Cantharellus formosus*. Still, *Cantharellus cibarius* can be found in conifer and hardwood habitats in other parts of North America and looks the same as *C. formosus* to the naked eye. Another similar but very large, yellow chanterelle, *Cantharellus californicus*, is found under oaks in California. *Cantharellus subalbidus*, the white chanterelle, often occurs at the same time as the Pacific golden chanterelle but is white and generally found at a higher elevation. Pacific golden chanterelles are mycorrhizal.

WHITE CHANTERELLE

(Cantharellus subalbidus)

The white chanterelle is a common and highly sought-after edible mushroom in the Pacific Northwest and Northern California. The white chanterelle can be quite large and dense. It is popular in specialty markets and restaurants. It is great in soup or as a sauce over meat, poultry, or pasta. Because of its high density, the white chanterelle does not release as much water as the Pacific golden chanterelle when cooked. Some people prefer eating the white chanterelle to the Pacific chanterelle. You will enjoy them both. Fundamentally, they are both fruity and appetizing.

It is also called the "giant white chanterelle."

Cap Characteristics

The cap is medium to large (generally three to six inches across), thick, dense, and colored white to cream. It is smooth, not scaly, and bruises yellow brown. The top of cap is sunken or depressed, funnel-shaped, and usually wavy.

Underside of Cap

The cap underside is white to cream in color, turning yellowish brown with age. Its blunt veins are sometimes interconnected.

Stalk

The stalk is generally one to four inches high, thick, and the same color as the cap. Its blunt veins or wrinkles run down the stalk and sometimes cross. The stalk also slightly tapers down from the cap and is solid, not hollow.

Fragrance and Edibility

The white chanterelle is fragrant and fruity, much like the Pacific golden chanterelle, with hints of apricot and pumpkin. It is very large, meaty, and delicious.

Habitat

You can find white chanterelles in the conifer forests of Oregon and Washington and Northern California. It fruits from October to December. Some report finding it under tanoak and manzanita in California. It is common in older conifer forests at high elevations. In the Cascade Mountain Range, the white chanterelle often occurs at the same time or a bit later than the Pacific golden chanterelle and often quite abundantly. The Cascade chanterelle (*Cantharellus cascadensis*) is also edible and looks similar in shape to the white chanterelle, but the cap is bright yellow, and it has a bulbous base.

The white chanterelle can fruit abundantly at high elevations across the Pacific Northwest.

YELLOW FOOT CHANTERELLE

(Craterellus tubaeformis)

The yellow foot chanterelle is a small edible chanterelle and is more delicate than the meatier Pacific golden or white chanterelle. It has a thin, hollow stem and is found in the winter and early spring in moist habitats. I included the yellow foot in *Fry, Thrive, or Die* because they are delicious! Sure, you will need to collect handfuls to make a meal out of these delicate and small specimens, but if you are on the Pacific Coast in the winter or spring, you are going to want to get to know and pick the yellow foot!

It is also referred to as "funnel chanterelle," "winter chanterelle," and *Cantharellus infundibuliformis.*

Cap Characteristics

The cap is small in size (generally one to two inches across), smooth, and colored brownish tan to yellowish orange. It is smooth, and the top of the cap is sunken or depressed.

Underside of Cap

The cap underside has white to yellowish blunt veins. The gills are widely spaced and sometimes interconnected.

Stalk

The stalk is generally one to three inches high and thin and hollow. Blunt veins or wrinkles run down the stalk.

Fragrance and Edibility

The yellow foot chanterelle is fruity and delicious. A favorite mushroom to eat, it is excellent in soup, with fish or poultry, and over rice. It should be cooked slowly to bring out its complex flavors.

Habitat

The yellow foot chanterelle is often abundant in moist environments in the winter and early spring. It is widely distributed under conifer forests, especially in Northern California and the Pacific Northwest. It commonly occurs in similar habitats and timing with hedgehog mushrooms.

BLACK TRUMPET

(Craterellus cornucopioides)

The black trumpet may look sinister, but it is delicious and well worth the effort to hunt and devour. It is one of my favorite edible mushrooms. Slow, gentle cooking brings out the flavors. It is great with chicken, fish, vegetables, and eggs. I hear from friends you can dry black trumpets and rehydrate them for later use. I've never done it (I eat them too fast), but I would love to try!

The black petunia shape and lack of gills or prominent veins make this mushroom easy to identify and perfect for the beginner to hunt and enjoy. It has no dangerous look-alike species to confuse it with. One of its common names is "trumpet of death," but whoever came up with this name should be banned from the forest. Don't believe it. It is one of the best edibles and finest fungal prizes in the West. You may hear it referred to as "horn of plenty" and "black chanterelle."

Cap Characteristics

The cap is generally medium-sized, two to four inches across. It is brownish if the conditions are dry and turns charcoal to black when wet. There is a central hollow from the center of the cap to the base of the stalk. The black trumpet resembles the shape of a petunia at maturity.

Underside of Cap

The cap underside is gray to brown. It is without gills or distinct veins and can be slightly wrinkled.

Stalk

The stalk is generally two to four inches high, thin, and leathery. It is gray, charcoal, or brown—not yellowish like chanterelles.

Fragrance

The black trumpet has a rich smoky flavor and fruity aroma.

Habitat

The black trumpet is commonly found during the winter, especially on the West Coast from Central California to Southern Oregon (although they can occasionally occur farther north). It fruits abundantly and in clusters under hardwoods, such as oak, tanoak, madrone, chinkapin, and beech. The black, gray, or brown color makes them difficult to spot in the shadows and among leaves. You find one, and then you realize you are standing in a patch of a hundred! The black trumpet is thought to be both a mycorrhizal and saprobic species (it can adapt to both). It grows on the forest floor and likes a dense layer of leaves, needles, and twigs. The mushroom is often coated with forest litter, so you may have to spend some time cleaning your black trumpets before you place them in your bag.

The black trumpet is very disguised—turning brown and tan if it is dry—and difficult to see on the forest floor.

FALSE CHANTERELLE

(Turbinellus floccosus)

The false chanterelle grows alongside many edible chanterelles, and if you are not paying attention to the chanterelles you are picking, it may find its way into your mushroom bag. If you don't catch it before it lands in your frying pan, you or your guests will likely experience gastric pains. The false chanterelle is distinguished by the prominent scales on its cap and the contrasting color between the reddish-orange cap and the white to pale-yellow stalk. The center of the false chanterelle cap and stalk is hollow. This mushroom should be avoided.

It is called by other names—"scaly chanterelle," "vase chanterelle"—and was formerly known as *Gomphus floccosus* until 2011.

Cap Characteristics

The cap is medium to large in size (generally three to five inches across), with prominent scales. It is vase-shaped and is red to orange on top. The top also is sunken or hollow.

Underside of Cap

The cap underside has white to yellowish blunt veins or wrinkles extending down the stalk.

Stalk

The stalk is generally two to four inches high and white to pale yellowish in color. Blunt veins or wrinkles run down the stalk.

Fragrance and Edibility

The false chanterelle has a mushroomy but not fruity smell. It causes gastric distress.

Habitat

The false chanterelle is widespread. You can find it before or during the Pacific golden and white chanterelle harvesting season. This mycorrhizal species is found in conifer forests across the West but is especially abundant in Northern California and the Pacific Northwest.

Similar Mushrooms That Should Be Avoided

Turbinellus floccosus, the false chanterelle described in this section, often grows in the same areas as the chanterelle and looks similar. You should avoid this mushroom because it can cause gastric distress. It can be distinguished from chanterelles by its scaly cap. The *Turbinellus* cap is also darker in color than the stem and not wavy.

The pale version of the false chanterelle is the *Gomphus bonari*, and the tan version is the *Gomphus koffmani*—both are not recommended as edibles. The pronounced scaliness of the cap and the difference in the red-orange color of the cap's top compared to the pale-whitish color beneath distinguishes the *Gomphus* and *Turbinellus* groups from the choice, edible chanterelles.

Another species that a beginner might confuse with a chanterelle and should not be consumed is the poisonous *Omphalotus olivascens,* or the western jack-o'-lantern mushroom. It is orange in color but has knife-edged gills (not blunt veins), and it lacks a wavy cap like a Pacific golden chanterelle. It is found in California. Several other orange *Omphalotus* species are also poisonous and, while not lethal, can cause severe gastric distress, vomiting, and diarrhea. The beginner might also confuse deadly *Cortinarius orellanus* and *Cortinarius speciosissimus* species that have orange to reddish brown caps and yellow to reddish brown gills, not folds like chanterelles.

Caring for and Preparing Chanterelles

Chanterelles can turn mushy and lose their flavor if they are washed with excessive amounts of water. If you clean your chanterelles well in the field, they won't require much washing. Chanterelles are generally bug-free. Prior to preparation, store your mushrooms in paper or cloth bags in a refrigerator. Dry-sautéing removes some of the water and concentrates the mushroom's flavor. To dry-sauté, first clean and slice the chanterelle. Put them in a skillet on high heat with *no* butter or oil for five minutes. Remove and place on a towel and let cool. Sautéed chanterelles can be used immediately or sealed in containers and frozen for later use. They are great in soups or as a sauce over meat, poultry, and pasta.

For you wild mushroom foodies, here are some other culinary opportunities:

- Mix into a wild mushroom risotto.
- Top a wild game dish like steelhead, duck, or venison.
- Chop, sauté, and incorporate into a hollandaise sauce.
- Slow-sauté and mix into a cheese omelet.
- Bake into a chicken pot pie.
- Add to gravy for an extra stylish sauce.
- Chop finely, sauté in garlic, and place mashed potatoes or white fish on top.
- Add to your favorite soup recipe.

And so many more options!

A MUSHROOM TRILOGY

Gordon Longhurst

I became hooked on foraging wild foods, especially mushrooms, one wild weekend during my college years in the early seventies. A buddy was building a ramshackle cabin in the coastal rainforests of Oregon, and a bunch of us went over to help him out. We ended up sharing one of the more memorable meals of my life. The feast began to take shape when one of the members of the party caught a four-pound steelhead from the Coquille River. That inspired us to go down on the beach at low tide and harvest buckets of mussels from the rocks along the shoreline. But the crowning glory was the discovery of chanterelles in a nearby woods. The spruce trees were tall, and their canopy created a Hansel and Gretel feeling, like a dark cathedral. The forest floor was a soft carpet of needles and scattered in profusion among the trees were bright-orange chanterelles. It looked like someone had chopped up Halloween pumpkins and flung the brilliant orange pieces everywhere. The mushroom soup, the trout steamed in ferns and lemons, and the mussels cooked in a rich broth of wine, butter, and garlic were fantastic. We were a happy band of hippies.

My friends Mel and Ann, who lived in the backcountry in Northern California, had survived a large forest fire the summer before my wife, Susan, and I went to visit them. It was spring now, and there were morels popping up everywhere among the burned trees, and we easily harvested shopping bags' worth. The four of us then set out to visit Charlotte and Patrick, who lived high in the hills of Mendocino County. Patrick worked at Fetzer Winery and had a great stash of premium wines. Ann had just received a large package of chocolates from her parents who lived in Switzerland. For the next couple of days, we ate stuffed morels, morel quiche, morel omelets, and washed them all down with wine that none of us could afford to buy. For dessert, we ate all the superb Swiss chocolates our piggy bodies could stand. As we dozed off on the deck watching the gorgeous sunset over Mendocino hills, our only worry was that it might be all downhill after this.

Years ago, living in the forest of Southern Oregon, we were fortunate to discover a perennial matsutake patch behind our house. At the time, the demand for matsutake was very high due to the Japanese market, and for several years, our young son and daughter made their Christmas spending money by picking and selling the "matsies" to wholesalers who were paying $50 a pound for top quality mushrooms. We all delighted in the treasure hunt, searching for the bumps in the layer of leaves under magnificent madrone trees and then shouting with pleasure as we uncovered the pungent mushrooms. We didn't mind selling them because we had found them disappointing as a culinary mushroom. Compared to so many other choice mushrooms we loved, they didn't measure up—until one day, when somebody clued me into grilling them on the barbecue with a little teriyaki sauce, and it was an epiphany. Suddenly we had a dilemma: sell 'em or eat 'em? We compromised and sold the high-value ones and scarfed down the rest. Life was good!

Matsutake motherlode.

Recipe:

CHANTERELLE SOUP WITH RED WINE AND BAGUETTE

INGREDIENTS:

2 tbs. olive oil
2 tbs. butter
2 white onions, sliced
4 garlic cloves, minced
10 oz. chanterelles, cut into 1 in. pieces
4 cups beef stock
1 cup red wine
1 bunch thyme
fresh baguette
salt and pepper

METHOD:

Melt the butter and add garlic and onion in a soup pan set to medium. Cook to soft and golden about for 10 minutes. Add the chanterelles and cook for another 7–10 minutes depending upon thickness. Stir in the beef stock, wine, and thyme. Season to taste with salt and pepper. Dip your baguette into the soup and enjoy.

2. MILK CAPS

Mushrooms bleed? Yes, the milk caps do. The bleeding milk cap and the delicious milk cap are edible, but frankly, they are not the best edible mushrooms. I can't tell you how many times I have collected them then stumbled upon chanterelles and boletes and thrown the milk caps out of my bag. I know it sounds harsh. Sometimes mushroom hunting can be cruel.

The color of the milk each one bleeds when cut distinguishes the three milk cap species included in this book: *red*—bleeding milk cap (*Lactarius rubrilacteus*); *carrot-colored*—delicious milk cap (*L. deliciosus*); and *white*—bearded milk cap (*L. torminosus*). It is important to note if the milk of any milk cap mushroom you collect is white or yellow, *it is not an edible milk cap* and should be avoided. The color of the latex can change over time when exposed to air, so take note of the color immediately after you cut the mushroom. Another distinguishing feature of milk caps is that their stalks break clean and straight like a piece of chalk.

White bleeding milk caps should not be consumed.

BLEEDING MILK CAP

(*Lactarius rubrilacteus*)

The bleeding milk cap exudes dark-red to purple milk when the gills are broken. This, combined with a green-staining cap and stalk, makes the bleeding milk cap distinctive.

It goes by other names—including "bloody milk cap," "red-bleeding milk cap," and *Lactarius sanguifluus*.

Cap Characteristics

The cap is medium-sized (generally two to five inches across), orangish to reddish brown, and contains an abundance of concentric rings at the top. Green stains streak the cap at maturity.

Underside of Cap

The cap underside has gills attached to the stem. They are reddish to orange brown and bleed a dark-red to purple latex when cut. It turns green when bruised (note—the color of the milk is very different from the color of a bruise).

Stalk

The stalk is generally two to four inches high and the same color as the cap, staining green when cut or bruised. The stalk breaks like chalk and is brittle, not stringy. There is no volva, ring, or veil.

Fragrance and Edibility

The bleeding milk cap has a nutty and a bit fruity fragrance. It is edible but not choice.

Habitat

Common in fall under conifer forests, especially in the West Coast, the bleeding milk cap is an ectomycorrhizal species that associates with conifers. It can fruit abundantly.

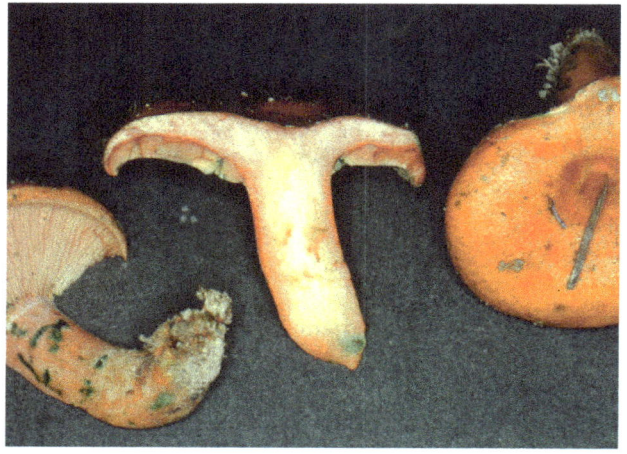

The bright-red bleeding and green staining are distinctive.

DELICIOUS MILK CAP

(Lactarius deliciosus)

The delicious milk cap has orange (carrot-colored) latex when the gills are broken or cut in combination with the green-staining cap when handled or injured; it makes this mushroom distinctive.

It goes by other names—including "saffron milk cap" and "orange juice milk cap."

Cap Characteristics

The cap is medium-sized (two to five inches across), often with concentric cap rings, and with a sunken center. It is orangish, orange brown, and can be grayish when young. When handled, the stains green at maturity.

Underside of Cap

The cap underside has gills attached to the stem. They are orange to orange brown and bleed a bright-orange latex when cut, turning green when bruised.

Stalk

The stalk is two to four inches high and is the same color as the cap, staining green when cut or bruised. The stalk breaks like chalk; it is brittle, not stringy. There is no volva, ring, or veil.

Fragrance and Edibility

The delicious milk cap has a nutty with a hint of pineapple fragrance. The taste is sometimes slightly peppery. It is edible but not choice in my opinion.

Habitat

A mycorrhizal fungus associating with conifers, the delicious milk cap is common in fall in conifer forests. It is widespread and can fruit abundantly in spruce and pine forests.

BEARDED MILK CAP

(Lactarius torminosus)

The bearded milk cap has a whitish to pale-pink to pinkish-orange cap that is hairy at the edges. It bleeds white latex and does not stain green. The cap center is sunken, and it has no volva or ring or veil. It is associated with birch and goes by other names such as "woolly milk cap" and "pink-fringed milk cap."

Cap Characteristics

The cap is medium-sized (generally two to four inches wide) and colored whitish to pale pink to pinkish orange. It contains an abundance of concentric rings at the top of the cap and is sunken in the center. The edge of the cap is rolled and hairy at the edge.

Underside of Cap

The cap underside has gills that are whitish to cream to light pink and bleed white. They are closely spaced and attached to the stem.

Stalk

The stalk is two to four inches high and the same color as the cap or slightly lighter. The stalk breaks like chalk, not stringy. There is no volva or veil.

Fragrance and Edibility

The bearded milk cap has a very peppery, acrid taste. It is not recommended as edible—it can cause gastric upset.

Habitat

Common in fall in conifer forests, especially associated with birch, the bearded milk cap is a mycorrhizal species and generally occurs in mixed birch forests or where birch is planted as an ornamental. It can fruit abundantly.

Similar Mushrooms That Should Be Avoided

Be careful to verify the red or orange color of the latex when the gills are cut or bruised. Don't eat a milk cap species if it bleeds yellow or white and has a very peppery or acrid taste. Other peppery milk caps that bleed white to yellow, like *Lactarius resimus,* should not be eaten.

Recipe:

MARINATED MILK CAPS

INGREDIENTS:

1 lb. bleeding or delicious milk caps—cleaned, trimmed,
and quartered or halved
1 tsp. salt
1/4 tsp. fresh ground black pepper
1/2 tbs. chopped wild thyme or
1/2 teaspoon dried thyme
3 cloves of garlic, sliced
1/4 cup oil (flavorless)
1/2 cup extra virgin olive oil
zest of a lemon, grated

METHOD:

Heat oil in a saucepan and add edible milk caps. Cook until it is light brown. Add garlic and cook until light brown. Stir in thyme, lemon zest, salt, and pepper. Remove mushrooms from the pan with a slotted spoon and put in a container. Top with the extra virgin olive oil, pressing the mushrooms down so that the oil covers them. Refrigerate. The marinated milk caps should be eaten within a couple of days of preparation. Spread them over your pasta sauce, pizza, or eggs.

2. HONEY MUSHROOM

(Armillaria mellea)

The honey mushroom is the largest and one of the longest-living life forms on the planet. Why do so few know anything about it? I suppose if it were a blue whale or a towering redwood and not a fungus, you would. Perhaps the honey mushroom needs an agent to work on its branding. The caps are actually pretty good to eat when they are young. But if you are a beginner, I would not recommend you eat it. It has white gills and a ring on the stem like a couple of deadly *Amanita* species. The honey mushroom is distinguished from *Amanita* by growing "almost always" in clusters or clumps of individual mushrooms from a concentrated base. But I have run into a solitary honey mushroom occasionally in the forest and in the yard.

The honey mushroom is part of the *Armillaria mellea* group and is a long-lived, white-rot fungus. It is both a destructive forest pathogen to conifer and hardwood trees as well as an edible addition to your culinary experience. *Armillaria mellea* encompasses a variable group of ten *Armillaria* mushroom species that have these general characteristics: presence of a yellow-edged ring on the upper stalk; a tough, fibrous stalk that tapers and fuses at the base; white spores; tiny hairs on the cap; and growth in tightly packed clusters at the base of infected trees and old stumps. The honey mushroom is also known as "honey fungus" and "oak root fungus."

A honey mushroom colony can be thousands of years old.

Cap Characteristics

The cap is medium to large in size (generally three to six inches across). The color is variable but generally shades of yellowish brown. It has numerous tiny hairs or scales on top.

Underside of Cap

The underside of the cap has gills attached to the stem. They are white to peach to yellowish in color and never turn brown.

Stalk

The stalk is generally long (five to ten inches high), tough, and the same color as the cap. There is no volva. The flesh is white to dingy yellow, and there's a stringy white pith in the stalk.

Fragrance and Edibility

The honey mushroom has a moderately sweet taste with a touch of bitter aftertaste. The mushroom cap is edible, but some people are intolerant of them. So eat a small amount the first time. Use them in any dish, just like you would use store-bought mushrooms. But unlike store-bought mushrooms, the stalks are very tough and fibrous and should be discarded. The tight clusters, with numerous individuals, can often fill baskets of mushrooms for your kitchen table.

Habitat

The honey mushroom is very common in a variety of habitats. This pathogen attacks both conifers and hardwoods with fungal colonies extending over many acres. In fact, it is earth's largest living organism (see the "Ancient Humongous Fungus" section). Fruiting occurs in the fall in dense clusters at the base of conifers, hardwoods, and buried wood. Sometimes it looks like honey mushrooms are growing on the ground, but with careful examination, you will find it is growing out of buried wood below the soil surface. It is common and prominent in forest areas across the Western United States.

Similar Mushrooms That Should Be Avoided

Toxic look-alikes include *Amanitas*—which, like the honey mushroom, also have white gills and a ring. *Amanitas*, however, have a volva at their base and free gills. To be certain a mushroom is the honey mushroom, make sure the clusters are growing from decayed wood and there is no volva at the base. There are also deadly *Cortinarius orellanus* and *Cortinarius speciosissimus* species that have orange to reddish brown caps and yellow to reddish brown gills.

THE ANCIENT HUMONGOUS FUNGUS AMONG US

Below the forest floor of the Blue Mountains looms a hidden whale of a creature. The discovery of this humongous life-form by forest pathologists Schmidt and Tatum (2008) made it the largest known living organism on earth—a title that was formerly claimed by the 110-foot-long, 200-ton blue whale. What is it? It's the *Armillaria ostoyae*, a honey mushroom fungus in the *Armillaria mellea* group. Found in a remote part of Eastern Oregon on the Malheur National Forest, it became the largest fungal colony and living organism in the world, spanning an area of 3.5 square miles (imagine 1,700 football fields placed side by side). This single organism is estimated to be somewhere between 2,400 and 8,600 years old and may be the oldest living organism on earth. The honey mushroom wears a variety of "caps." If you are a tree, it's a pathogen and a deadly killer; if you are a cavity-nesting wildlife species living in a decaying tree, it is a place to call home; and if you are a foraging mushroom hunter, it is dinner.

The team of forest pathologists discovered the ancient fungus and set out to map the extent of its reach. The team grew fungal samples in petri dishes to see if they fused together, an indication that they were from the same genetic individual. Utilizing DNA fingerprinting, they determined where one individual fungus ended and another began. What they found was that a single genetic fungal colony weighed approximately 35,000 tons—or 175 times the weight of a blue whale.

Most trees benefit from mycorrhizal fungi and mushrooms. The *Armillaria mellea* group of honey mushrooms, however, are pathogens that suck the life out of a variety of tree species. The mycelium of the honey mushroom girdles the trunk of the tree, killing all living tissue. It's a slow-rotting death that can take decades to complete, but when the fungus is finished, the tree has been recycled back into the soil for the benefit of future forests. The fungus then spreads to uninfected parts of the forest at a glacial speed of one to three feet a year.

Recipe:

SAUTÉED HONEY MUSHROOM CAPS

INGREDIENTS:

1 lb. fresh, young honey mushroom caps (discard the stems)
salt and pepper
2 tbs. grapeseed or canola oil
2 tbs. butter

METHOD:

In a frying pan, heat oil on medium high. Cook caps for 5 minutes until they are light-colored. Add butter and continue to cook the caps for another 5 minutes until mushrooms are golden and caramelized. Make sure your honey caps are thoroughly cooked. Uncooked honey mushrooms can cause an upset stomach. Season with salt and pepper to taste.

4. AMANITAS

The *Amanita* group contains some of the most beautiful, deadliest, and most hallucinogenic species of mushrooms. The *Amanita* group includes some of the most delicious edibles too. But differentiating between edible and poisonous species requires a keen eye and considerable field experience. I have not included edible *Amanitas* in this pocket guide because of the difficulty in distinguishing them for inexperienced collectors and for personal reasons—I don't want you to get sick or die if you misidentify an *Amanita* mushroom. Slipping a deadly *Amanita* into a meal has been involved with treacherous murders for millennia, including Roman Emperor Claudius AD 54 and Holy Roman Emperor Charles VI in 1740. I don't want you to end up like Claude or Chuck.

Two species in this section, the death cap and destroying angel, are deadly poisonous. The other two species in the section, the fly agaric and panther amanita, are hallucinogenic and can cause severe gastric distress. There is no harm in handling *Amanita* mushrooms; they must be ingested to be poisonous.

Incidentally, the fly agaric in this section is the poster child for mushrooms, video games, and trinkets all over the world. The classic red cap with white spots is iconic in the marketplace. But is there a connection to Christmas traditions?

The *Amanita* species, as a general group, have white to very pale yellow gills, and when young, they have a universal veil that encases the mushroom like an egg. As the mushroom grows, the universal veil ruptures, leaving a sac at the base called the volva. The volva is a key feature of *Amanitas*, so it is important to always dig beneath the soil surface of white and free-gilled mushrooms to determine if a volva is present. Another important feature is that the veil protects the gills and leaves a ring near the top of the stalk. *Amanitas* have gills that are "free" and not attached to the stalk. You should never eat a white-gilled mushroom until you know the deadly *Amanita* species in all their forms.

The fly agaric.

DEATH CAP

(Amanita phalloides)

The death cap is the number 1 killer from mushroom poisoning.

The death cap name says it all. It is beautiful to look at, but if ingested, it is deadly. It is also known as "deadly amanita" and "stinking amanita."

Cap Characteristics

The cap is medium-sized (generally three to six inches across) and colored pale green, yellowish green, to olive brown. Its color can fade with age or intense rain. The cap has very scattered white warts or flakes, which are remnants of the universal veil. The cap tends to be bald (without warts) with age, and radial lines are absent at the edge.

Underside of Cap

The cap underside has white gills that stay white at maturity; they are free from the stem.

Stalk

The stalk is generally three to five inches long and colored white, sometimes with a hint of metallic green. It has a volva. The veil initially covering the gills evolves to a white to yellowish skirtlike veil on the stalk. Weather or intense rain can wash off the ring.

Fragrance and Edibility

AVOID CONSUMPTION. DEADLY. It smells pungent and metallic. If ingested, a single mushroom can kill several people. It is the number 1 cause of fatal mushroom poisonings worldwide. Age and weather can alter the smell of the death cap.

Habitat

This mycorrhizal species only occurs in association with trees. Tree roots can extend into lawn areas, so the death cap is sometimes found in lawns or turfs. It was originally found in Europe but has proved to be highly mobile. With the planting of non-native trees with roots colonized with *A. phalloides*, it is now present in North America, Australia, and South America. It is widespread across the Western United States, occurring frequently under oaks and various hardwood species and with a wide variety of popular ornamentals.

DESTROYING ANGEL

(Amanita ocreata)

The destroying angel is deadly poisonous.

The destroying angel is an appropriate name for the deadly *Amanita ocreata*. The destroying angel is also known as "death angel" and "angel of death." Notice a theme here?

Cap Characteristics

The cap is medium-sized, generally three to six inches across. It is white at the edges and white, tan, yellowish, or buff at the center. The cap tends to be bald but can have a thin patch, a remnant of the universal veil. Radial lines generally absent at the edge.

Underside of Cap

The cap underside has white gills that stay white at maturity; they are free from the stem.

Stalk

The stalk is generally three to five inches long and colored white, with a volva at the base. It has a universal veil initially covering the gills and evolving to a white to yellowish skirtlike veil high on the stalk.

Fragrance and Edibility

AVOID CONSUMPTION. DEADLY POISONOUS. It has an unpleasant, sickly odor that may be too faint to detect on a windy day. If ingested, one mushroom can kill. It is one of the major causes of fatal mushroom poisonings worldwide.

Habitat

The destroying angel is a mycorrhizal fungi that only occurs in association with trees. It is widespread in California, Oregon, and Washington, and it is commonly found under oaks and hazelnuts in California. It generally fruits in the winter and spring. Another poisonous white *Amanita*, *Amanita smithiana* is also found in the Pacific Northwest.

White *Amanitas* are sometimes confused with the common and widespread *Agaricus* species, such as the horse mushroom (*Agaricus arvensis*) and meadow mushroom (*Agaricus campestris*). These two *Agaricus* species have a white stem and cap, but unlike *Amanita* species, they have chocolate-brown gills and spore prints with maturity. *Leucoagaricus naucinus*, the woman on horseback mushroom, resembles the destroying angel in color and size and has free white gills; however, *Leucoagaricus* species have no volva, which is why it is very important to dig up every white-gilled mushroom you collect to inspect for the presence of a volva.

Amanita species are sometimes confused with the edible rose-gilled grisette (*Volvariella speciosa*), a popular edible with some Asian cultures. The rose-gilled grisette also has a white stem, a cap, and a volva, and it can look much like an *Amanita*. However, the gills turn a dull reddish to light pinkish with age. It is better to avoid this edible *Volvariella* until you are experienced with identifying toxic *Amanita* species. The rose-gilled grisette occurs on all continents except Antarctica. Recent DNA evidence has reclassified *Volvariella speciosa* as *Volvopluteus gloiocephalus*.

THE FLY AGARIC

(Amanita muscaria)

The fly agaric is the mushroom of legend and lore. It has both a complex chemistry and history.

The iconic fly agaric is a true cosmopolitan species that over the last hundred years has spread across the world, hitchhiking as mycorrhizae on the roots of both conifers and hardwoods. I've seen fly agarics in Central America, Australia, New Zealand, Thailand, and China. Native to the Northern Hemisphere and far Northern climates, the "classic" bright-red cap with white warts spotting the cap has left a mark on many cultures. Gnomes seem to love this toadstool, and you can see statues and trinkets of them together in garden stores from Oregon to Norway to Thailand. Children of all ages use the mushroom to "power up" in the *Mario Brothers* video games. Dogs dress up like the fly agaric for Halloween. No kidding.

But this mushroom goes further back in history. Legend has it that the Viking berserkers ingested the hallucinogenic mushroom to be fearless in battle (Fartur 2019). There is also evidence to believe that the fly agaric inspired the legend of Santa Claus (Main 2013). And this is just the tip of the fly agaric iceberg.

This mushroom is also known as "fly amanita." The common names have the word *fly* in them because *A. muscaria* has been used for years to control flies (Lumpert and Kreft 2016) in places like Siberia and Slovenia. Pieces of mushroom placed in milk on the windowsill attract flies, which upon ingestion fly off to their death in an intoxicated state.

Cap Characteristics

The cap is medium to large in size (three to eight inches across) and colored bright red to reddish orange. It has scattered white warts or flakes, which are remnants of the universal veil. The cap can fade with age or intense rain, and radial lines are common at the edge.

Underside of Cap

The cap underside has white gills that stay white at maturity; they are free from the stem.

Stalk

The stalk is generally three to six inches long and colored white. The base is bulbous and with a volva. It has a partial veil initially covering the gills and evolving to a white to yellowish skirt-like veil on the stalk. One or more scaly rings can occur at the bulbous base. The lower end of the stalk can turn a dingy yellow at the bulbous base.

Fragrance and Edibility

The fly agaric has a mild earthy smell. Although poisonous, ingestion does not result in death. *Amanita muscaria* is noted for its hallucinogenic properties; the responsible chemicals are the neurotoxins ibotenic acid and muscimol.

Habitat

The fly agaric is widely distributed globally. In the Western United States, it forms a mycorrhizal association with conifer and hardwood trees. The fly agaric is an opportunist and can be found in forests, parks, and urban settings. It is often growing abundantly in clusters and sometimes in circles, known as fairy rings.

PANTHER AMANITA

(Amanita pantherina)

The panther amanita is common and widely distributed across the Western Unite States.

The panther amanita resembles the fly agaric. It has a volva, ring present at the bulbous base, free gills, and warts on the cap. It differs in that it has a tan or dark-brown cap and can fruit in the spring. This mushroom is also known as "panther cap" and "false blusher."

Cap Characteristics

The cap is medium to large, generally three to eight inches across. In coloring, it is tan to dark brown, never red or reddish orange. It has scattered white warts or flakes, which are remnants of the universal veil, and radial lines are found at the edge.

Sometimes the warts on top of the caps of *Amanitas* will wash off in a heavy rain. Be careful to check for a volva at the base.

Underside of Cap

The cap underside has white gills that stay white at maturity; they are free from the stalk.

Stalk

The stalk is generally two to five inches long, colored white, and does not blush red when bruised. The bulbous base has a volva. It has a partial veil initially covering the gills and evolving to a skirtlike veil on the stalk. One or more scaly rings can occur at the bulbous base.

Fragrance and Edibility

POISONOUS. IT SHOULD BE AVOIDED. The panther amanita causes nausea following ingestion. It intergrades with several other *Amanita* species that are tan and brown. It does not contain the deadly amatoxin, but some look-alikes do. The true *Amanita pantherina* contains the psychoactive compounds ibotenic acid and muscimol and is used as a psychoactive substance less often than its distinguishable relative *Amanita muscaria*. The average psychoactive potency of the *A. pantherina* is unknown. The panther amanita has a mild smell when intact, but if the flesh is squeezed, it smells like radish or raw potatoes.

Habitat

A mycorrhizal fungus, the panther amanita can be found growing under hardwood and conifer trees. It is widely distributed in the Western United States and fruits in both fall and spring.

Similar Mushrooms That Should Be Avoided

Highly toxic look-alikes containing amatoxins include *Amanita verna, Amanita smithiana,* and *Amanita bisporagera.* The real danger in eating the panther amanita or fly agaric is mistaking it for a deadly amanita, such as the destroying angel or death cap. You only make that mistake once. A single mushroom can kill!

The panther amanita is similar to *Amanita rubescens* (called blushing amanita or "the blusher"), but the panther amanita doesn't "blush" red or pink when the flesh is damaged, hence the origin of one of its common names—the "false blusher." Blushing is a key feature in differentiating these species. There are differing reports about the edibility of *Amanita rubescens*, but because of its similarity to other toxic *Amanitas*, it should be avoided.

MURDER

Claudius was a Roman emperor (10 BC–AD 54) thought to have been murdered through mushroom poisoning (Marmion 2002). Legend has it that a deadly *Amanita* was added to his plate of mushrooms. Who was the killer? Claudius had many enemies. For starters, he had executed his third wife, Messalina, for adultery and married his niece Agrippina. The most likely suspects appear to be either Agrippina or her friend Locusta. But why? With Claudius's death, Agrippina's nephew Nero would ascend to become emperor, placing this as one of the greatest culinary crimes that would change the course of history.

Mushrooms were considered a delicacy for many Roman elites at that time. In fact, one *Amanita* species was highly prized by a host of Roman emperors and is called Caesar's amanita (*Amanita caesarea*). It is widely distributed in the Western United States and Southern Europe. The Caesar's amanita should be avoided unless you really know the *Amanita* group because several similar-looking *Amanita* species are deadly poisonous. Some of the *Amanitas* you spot in the woods of the Western United States may be one of the *Amanitas* added to Emperor Claudius's plate the night he died—the destroying angel (*Amanita ocreata*) or the death cap (*Amanita phalloides*).

The death cap packs a mighty poisonous punch. One medium-sized mushroom can kill three to four people. Today, the death cap and the destroying angel cause most mushroom poisoning deaths. About forty people a year suffer from severe mushroom poisoning in the United States, with approximately three deaths per year. Once ingested, the death cap and destroying angel cause diarrhea, nausea, vomiting, and with time, liver and kidney failure.

Asian Americans are the most frequently poisoned by the death cap because it can be mistaken for a popular and edible Asian mushroom called the paddy straw mushroom (*Volvariella vol-*

vacea). The amatoxin in the death cap and destroying angel is not affected by cooking, so sautéing, baking, or broiling does nothing to prevent the impact of ingestion. It is interesting to note that the death cap is not native to North America. It most likely arrived on trees imported from Europe in the early twentieth century. Today it is widespread.

I've gathered and have eaten the edible Caesar's amanita. I was careful to identify all the diagnostic characteristics of each mushroom that was cooked. The first time I found some, I was with seven of my mushroom hunting friends, and we were hunting mushrooms up the North Umpqua River outside of Roseburg, Oregon. The cabin we were staying at was near a campground. The large, orange-red caps lining the margin of the campground were easy to spot: free gills, a volva, and a smooth, dark-orange to orange-red cap lacking warts. I thought, "My first Caesar's amanita!" But did I have the guts to cook them and eat them with friends over the steelhead salmon a friend had caught for dinner? The Caesar's amanitas had an orange-red cap (darker orange red toward the center of the cap), the cap surface was smooth, no scales, cap margins were conspicuously striated, and free gills were pale golden yellow. Even with the careful observations and decades of hunting and eating mushrooms, I was, frankly, a bit nervous about eating the Caesar's amanita. The Caesar's amanita was delicious with the fish. But would I do it again?

ANCIENT HALLUCINOGEN

The fly agaric has a long history of use as an intoxicant and entheogen by indigenous people.

There is considerable documentation that the fly agaric mushroom was used as an intoxicant and entheogen by the indigenous peoples of Siberia and the Sámi, in the northern regions of Norway, Finland, and Sweden. The usage in Siberia seemed to be for ceremonial purposes (Wasson 1979, Feeney 2022). There has been much speculation on the historical use of this mushroom by other cultures, including the Viking berserkers (Fartur 2019).

Reported methods for how to weaken the toxicity and the hallucinogenic effects of *A. muscaria* include parboiling the mushroom twice with water. Others report that drying, then soaking in lemon juice, milk, or yogurt can reduce the nausea and uncomfortable reaction following ingestion. The fly agaric continues to be used ceremoniously in northern parts of Europe and Asia by the native people.

AMANITA MUSCARIA AND CHRISTMAS TRADITIONS?

It seems strange, when you think about it, that so many of our Christmas traditions have an unusual story line—flying reindeer, a tree with red ornaments and packages, a laughing man dressed in a red coat with white buttons making his epic trip from the North Pole every year. We celebrate each winter by stringing colored decorations on a conifer tree, then we wait for the jolly old man to descend down the chimney to deliver us our gifts—gifts wrapped in red and white delivered from a flying sleigh! How did our holiday customs come to be? We could write Santa a letter and ask him directly, but let's save the postage and dig a little deeper into the historical references to this legendary fungus, the fly agaric.

Saint Nicholas did not start out being Santa Claus. He was a fourth-century saint from Greece known for his generosity and gift-giving. It wasn't until the nineteenth century that the poet Clement Clark Moore wrote a poem that helped transform the good saint into his modern image. In Moore's 1822 poem, "A Visit from St. Nicholas," he combined elements of Medieval

European Christianity, Eastern Orthodox Church, and Dutch folklore to create the mythical poem. Moore, perhaps unwittingly, left clues by referencing ancient Nordic culture and shamanic tradition that would become an iconic symbol of modern Christmas: the eight flying reindeer.

"Whoever heard of reindeer flying, except as shamanic vehicles?" asks Carl Ruck, a professor at Boston University and expert on rituals. He continues, "It is well established reindeers have fondness for [*Amanita muscaria*] mushrooms." Research has shown that reindeer do seek out the *Amanita muscaria*.

Deputy Editor of the *Pharmaceutical Journal* Andrew Haynes wrote in 2010 that animals deliberately seek out the red-and-white-spotted mushroom in their habitats, and they seem to experience altered states of behavior. For humans, a common side effect of the fly agaric mushrooms is the feeling of flying, so it's interesting the legend about Santa's reindeer is that they can fly. Perhaps in extreme Northern climates, humans and reindeer eat the fly agaric to escape the monotony of long, dreary winters.

The winter solstice has been a time of festivity and celebration for thousands of years with dancing, music, meals, and gatherings long before Christmas celebrations in traditional Christian sense. *The Huffington Post* (December 16, 2016) devotes a detailed description regarding the magical history of Yule and pagan winter solstice celebrations. This same *Huffington* post article chronicles that pagan cultures of Central Europe, for example, celebrated the Yule midwinter festival around the winter solstice. The ancient Romans held the festival of Saturnalia in late December in honor of the god Saturn. But of greater fungal interest is a practice in Siberian and Arctic regions (a suburb of the North Pole), where in December, shamans drop into the locals' homes, giving away presents of vision-inducing mushrooms.

Doug Main describes eight ways that magic mushrooms explain the Santa Story (Live Science 2013). His article raises the speculation: Is Santa the modern equivalent of a shaman who ceremonially uses certain mind-altering plants and fungi to commune with a natural world of wonder? Real accounts indicate that in recent history, shamans or priests—who were connected to the pre-Christian traditions—would collect *Amanita muscaria*, dry them, and then give them as gifts on the winter solstice. Drying the *Amanita muscaria* was traditionally done by hanging the red-and-white mushroom from conifer trees. It is interesting that drying *Amanita muscaria* helps reduce the impact of the nauseating effect of ibotenic acid while preserving the mild-altering effects of muscimol (Feeney 2010).

Dr. Carl Ruck, a Boston College professor interviewed in *Newsweek* (December 24, 2012), believes "that perhaps Siberians who ingested the mushrooms hallucinated that the grazing reindeer were flying? At first glance, one thinks it's ridiculous, but it's not. Whoever heard of reindeer flying? I think it's becoming general knowledge that Santa is taking a trip with his reindeer." He continues, "The Christmas tree is a motif that you find in Nordic mythology of Christmas. It has to do with the solstice; gifts under the tree might well be a reference to the way the mushroom grows around the sacred tree. The red and white mushrooms are 'gifts' found under a conifer tree."

Some scholars refute the Santa–*Amanita muscaria* roots of Christmas. It is, without question, an unprovable hypothesis, and it entirely depends upon your propensity to speculate. But perhaps it would be refreshing to have the Christmas holiday experience be a time of wonder, self-reflection, healing, and contemplation instead of the commercial bombardment that it has become today.

AMANITA TOXINS

Know your mushrooms before you ingest any mushroom with white gills that do not attach to the stem (free gills).

The destroying angel and death cap contain amatoxins (along with *Amanita smithiana*, *Amanita verna*, and *Amanita bisporagera* and perhaps a few others). Amatoxins are no joke; they cause severe liver and renal failure. Severe reactions occur many hours after ingestion, so by the time you start feeling sick, the amatoxins have already been absorbed into your system, and you don't have many survival options. Amatoxins are also present in some deadly *Galerina* and *Lepiota* (e.g., *Lepiota brunneoincarnata*) species, so know your mushrooms before you eat any!

The fly agaric and the panther amanita do not contain amatoxins. They contain ibotenic acid and muscimol that affect mental perception but do not attack the liver and renal tissue. Ibotenic acid has side effects that may cause nausea and uncomfortableness. Muscimol is considered the chemical with the primary mind-altering properties (Feeney 2010). Some ibotenic acid can convert to muscimol by drying to a temperature between 165°F and 185°F. Adding dried fly agarics to lemon juice or fermenting in milk or yogurt is also said to help bacteria detoxify the fly agaric (Tsunoda 1993).

5. PARASOLS

The parasols are big, majestic, and easy to spot. The parasols are a large family that include delicious as well as poisonous species. I have included three important parasol mushrooms in this guide—one simply called the parasol (which is a culinary favorite), the shaggy parasol (which is very good but for a few people causes gastrointestinal side effects), and one called the false parasol (which is poisonous and must be avoided).

Common to all parasols is the presence of a veil covering the young gills and forming a ring on the stock. While you may confuse this characteristic feature with *Amanitas*, the parasols do not have a volva around the base of the stem, which is why it is always important to carefully dig up the stalk of mushrooms you are considering for consumption to check for a volva. The most common characteristic differentiating the parasol and shaggy parasol from the poisonous false parasol is the color of the gills. The parasol and the shaggy parasol have white gills, while the false parasol has greenish gills or a rare green spore print. If you are considering consuming a parasol mushroom, you will want to take a spore print of your specimens to see if the print is white or green.

PARASOL

(Macrolepiota procera)

One of the most delicious of all mushrooms, the parasol is tall with a large, thin cap. The flesh of this stately species does not turn orange or red when rubbed or cut. The parasol mushroom goes by other names such as *Lepiota procera* and *Agaricus procerus*.

Cap Characteristics

The cap is medium to large in size (three to eight inches). It is drumstick-shaped when young and flattens at maturity. Smooth and brown at first, with age it forms shaggy scales on a tan to grayish cap. The center of the cap remains brown.

Underside of Cap

The cap underside has white gills that stay white at maturity, not greenish; they are free from the stem. A veil is present when it is young.

Stalk

The stalk is tall (generally six to twelve inches high), thin (less than 3/8 inch thick), and has a prominent double-edged ring on the stalk. The base is without a volva. The flesh does not stain red or orange when rubbed or bruised. The stalk surface is covered with delicate brown scales.

Fragrance and Edibility

The parasol mushroom has a nutty sweet aroma that some people say remind them of maple syrup. It is one of the best wild edibles. The caps are delicious when sautéed in butter or coated with bread crumbs and fried. Use it in any dish just like store-bought mushrooms. But unlike store-bought mushroom, the stalks are tough and fibrous and should be discarded.

Habitat

Often abundant, the parasol can be found growing near conifers, roads, pastures, parks, and grassy areas. You can find them in the summer and fall.

SHAGGY PARASOL

(Chlorophyllum rhacodes)

The shaggy parasol is thicker and shorter than the parasol mushroom. The flesh of the shaggy parasol turns orange or reddish when bruised or cut. I find it delicious, but some people get an upset stomach after eating. The shaggy parasol goes by other names such as "leppie" and *Lepiota rhacodes*.

Cap Characteristics

The cap is medium to large in size (three to seven inches). It is egg-shaped when young, flattening somewhat with maturity. Smooth and brown at first, it forms prominent shaggy scales on a tan to grayish background with age. The center of the cap remains brown.

Underside of Cap

The cap underside has white gills that stay white at maturity, not greenish; they are free from the stalk. A veil is present when it is young.

Stalk

The stalk is short (generally two to four inches high) and about a half inch thick. It is shorter and thicker than the parasol mushroom. The bulbous base is without a volva. The flesh stains orange, red, or maroon when rubbed or bruised. The stalk is without scales and has a prominent double-edged ring.

Fragrance and Edibility

The shaggy parasol has a mild earthy, pleasant fragrance. Many people eat the shaggy parasol, which is generally considered a delicacy; however, some people have a gastrointestinal reaction. For first-time consumption, eat only a small amount, and cook it on high heat to reduce the potential for an adverse side effects.

Habitat

Often in clusters, the shaggy parasol can be found growing abundantly anywhere—woods, orchards, roadsides, pastures, parks, bark piles, and grassy areas. It can be found in spring, summer, and fall.

FALSE PARASOL

(Chlorophyllum molybdites)

Eating the false parasol is a major source of mushroom poisonings in the Western United States. It looks like the parasol mushroom but has distinctive greenish gills and a green spore print. The false parasol is also known as "green-spored *Lepiota*," "Morgan's lepiota," and *Lepiota molybdite*. It is also aptly called "the vomiter."

Cap Characteristics

The cap is medium to large in size (four to twelve inches). In coloring, it is whitish to pinkish brown with coarse, brown scales. It is drumstick-shaped when young but flattens with maturity.

Underside of Cap

The cap underside has white gills that turn dark and green at maturity. It has a rare green spore print, and the gills are free from the stem.

The false parasol looks a lot like the parasol but has green spores and green spore print.

Stalk

The stalk is four to twelve inches tall and slender to medium in thickness. It has as a double-edged ring. The stalk is without scales, and the base is swollen but not bulbous. It has no volva. The flesh is white but sometimes stains orange or reddish brown.

Fragrance and Edibility

IT IS POISONOUS. DO NOT CONSUME. The false parasol has an earthy smell. It causes vomiting and diarrhea one to three hours after consumption. The false parasol is conspicuous in lawns and urban areas, and for this reason, it is a major source of food poisoning because it can be mistaken for the parasol, shaggy parasol, and shaggy mane mushroom. Poisonings can be severe, though no deaths have been documented.

Habitat

Often in clusters and gregarious, the false parasol is common in lawns and pastures but can be found anywhere. It fruits in the spring, summer, and fall and thrives in hot, dry weather.

Similar Mushrooms That Should Be Avoided

The *Lepiota subincarnata* is a poisonous *Lepiota* that contains deadly amatoxins. The cap of the *Lepiota subincarnata* is light red to red brown and cream in color closer to the margin. The gills are whitish, and the flesh is white to pinkish toward the top. The stem may be slightly larger at the base, cream-colored with patches of the cap color. The odor is somewhat fruity, and the taste is unpleasant. It has been documented to grow in lawns and parks in North America. Toxic look-alikes also include some *Amanita* species. They also have an ornamented cap like the parasol mushroom and have white free gills at maturity. The *Amanita* species have a volva at their base; the parasols do not.

Recipe:

BREADED PARASOL

Try this breaded and fried parasol recipe. Parasols are also good sautéed, baked, and as an addition to soups.

INGREDIENTS:

1 egg
2 oz. flour
1/2 cup bread crumbs
frying oil
salt, as preferred
1/2 lemon

METHOD:

Clean the parasols without water. Remove and discard the stems. Whisk egg with a pinch of salt and pour into a plate. Add flour to a second plate. Coat mushrooms with the egg mixture, then in the flour. Fry in oil until they develop a crisp golden texture. This will take around 2 minutes on each side of the cap. Serve with a squeeze of lemon.

6. AMERICAN MATSUTAKE

(Tricholoma murrillianum)

The American matsutake—revered by the Japanese—is sweet, spicy, and . . . pricy. Don't sauté it; it ruins the complex matsutake flavor! Baked or grilled is the way to savor. The American matsutake (*Tricholoma murrillianum*) has a distinctive spicy and cinnamon smell. The stem and cap separate in thin, stringy sections and have a cinnamon color with age. The gills are notched at the attachment with the stem. Be careful not to confuse it with poisonous *Amanita* species that, unlike matsutake, have a volva and free gills.

The American matsutake is also known as "matsutake," "matsi," "pine mushroom," and *Armillaria matsutake*.

Cap Characteristics

The cap is medium to large in size (two to nine inches). Its margin is in rolled at first, upturned with age. The surface is dry white to light yellow, bruising a light cinnamon color with age. The flesh breaks into stringy pieces.

Underside of Cap

The cap underside has white and crowded gills that turn a light cinnamon color with age. They are notched at their attachment to the stem.

Stalk

The stalk is two to six inches long and medium in thickness—greater than a half inch. It is tough and stringy and has a cottony veil when young that persists as a ring on the stalk.

Fragrance and Edibility

American matsutake smells like a combination of wet earth, sweet spice, cinnamon, and pines. It's very distinctive. The unique and complex flavor of the matsutake is lost by sautéing. It is better baked, grilled, or roasted. The Japanese have revered this mushroom for over a millennium for its complicated smell and flavor (and supposed aphrodisiac qualities). Young, unopened specimens can sell for several hundred dollars a pound during some holidays in Japan. The Japanese matsutake *Tricholoma magnivelare* is very similar to the American matsutake. Lots of American matsutake get shipped to Japan where demand exceeds domestic production. Since the matsutake has a stringy texture, you can pull thin, one-eighth-inch strips from the mushroom stem and cap. Grill or roast it till it turns light brown. I've seen matsutake eaten raw on top of salads in Japan, but I'm not sure if it would please a typical American palate. The matsutake is chewy but has a unique taste that lingers on your senses. I often add a splash of soy or teriyaki sauce when serving matsutake with meat or other vegetables.

The matsutake is a desirable and expensive commodity in Asian markets.

Habitat

DNA evidence presented in 2017 indicates there are three matsutake species in North America: one in the Eastern United

States (*Tricholoma magnivelare*); one in the Western United States (*Tricholoma murrillianum*); and one species from Mexico (*Tricholoma mesoamericanum*). The American matsutake is widely distributed in conifer forests in the Cascades and Rocky Mountains. It also occurs frequently with hardwoods like madrone, manzanita, chinquapin, and tanoak along the California and Oregon coastal mountains. American matsutake tends to fruit in sandy, well-drained soils and fruits from August to February in the Western United States, depending upon climate and elevation.

The American matsutake fruit in patches, known as "*shiros*," year after year in the same location.

During the 1980s and 1990s, American matsutake prices often peaked, creating a gold rush mentality in Western forests (Amaranthus et al. 1998). Makeshift camp communities and buyers would flock to harvest areas. Conflicts over prime fruiting patches sometimes occurred, and it was typical for harvesters to carry guns for protection and fire shots in the air to warn others to stay away. Concerns regarding over harvesting matsutake was high, but research showed that careful excavation of the mushroom without raking the soil surface had no impact on matsutake production (Amaranthus et al. 2000).

Commercial harvest of matsutake had a gold rush fervor when prices soared to $100–$200 per pound.

Similar Mushrooms That Should Be Avoided

Some poisonous *Amanita* species are also white but have a volva at the base beneath the soil surface and free gills. I have seen the destroying angel *Amanita* fruit a foot away from a cluster of American matsutake. Clearly, it is important to carefully inspect every mushroom while you pick and again before you eat it.

Fun Facts about *Matsutake*

- The matsutake is ectomycorrhizal, with a wide variety of conifers and hardwoods. But on the microscope, I have frequently seen the mutualistic mycorrhizal relationship turn into a saprophytic relationship when the host tree begins to decline. The matsutake fungus then decomposes the fine roots.

- In the Japanese language, *matsu* means "pine" and *take* means "mushroom." The matsutake is closely associated with the red pine tree in Japan.

- For over a thousand years, the matsutake has had special meaning in Japanese culture. In the Imperial Court in Kyoto, the matsutake was so sacred that women were not allowed to utter the word. Large, young, and unopened specimens in Japanese markets can fetch several hundred dollars a pound. The matsutake is not only considered as a symbol of health and happiness but also has a phallic shape when young that is believed to be an aphrodisiac when consumed.

- *Allotropa virgata*, also known as candy cane, is a beautiful candy-stripped achlorophyllous plant that indicates matsutake mycelium in the soil. The candy cane root feeds on the matsutake mycelium, taking the nutrients it needs. The candy cane stem emerging from the forest floor is a reliable indicator that the fungal mycelium of matsutake is present, and the opportunity for a matsutake mushrooms occurs when moisture and temperatures are good for fruiting.

Allotropa virgata, the candy cane plant, is both beautiful and a useful indicator of the presence of American matsutake in the Cascade Mountain Range.

THE JAPANESE AMERICAN MATSUTAKE EXPERIENCE

Eric Ballinger

My first memories of matsutake were sitting at the countertop of my grandparents' kitchen, sipping the *dashi* broth of my Grandmother Kazuko's matsutake soup. The curls of steam lifted the scent of this mushroom to all corners of their home. My grandparents always seemed to have an abundance of these fragrant fungi, but to what extent, I had no idea. And where did they come from?

In the Japanese American community of the Columbia Gorge of Oregon, every family had their special culinary contribution.

My grandparents' specialty was picking, preparing, and sharing matsutake mushrooms. My grandmother would make *nishime*, a matsutake stew—combined with *satoimo*, *kamaboko*, *konnyaku*, chicken, carrots, and *daikon*. And for the New Year's celebration *Mochitsuki*, she would make *matsutake gohan* to share with the Toda and Ogawa families on the other side of the Columbia River. The rustic recipes were brought from the old country.

Every fall, my grandparents would return to the rain-soaked woods of the Cascades. Their tools were keen eyes, patience, red-handled wooden dowels, a cloth rice bag, a knife, a brush, and a curious *yonsei* ("fourth-generation grandson")—me. We would clean the matsutake mushrooms on the back of the covered porch and prepare them for a bus ride to Ontario, Oregon, for the Hinatsu family. In return, the Hinatsus would send back a hundred-pound bag of onions and potatoes they grew on their farm. Perhaps a freshly caught salmon would appear as word got out that my grandparents had returned from the woods. It could be a box of perfect yellow Asian pears from Shig Imai's orchard.

Matsutake mushrooms have been used in Japan since the Neolithic period (8000–3000 BCE) in the final period of the Stone Age. Within my own family in Oregon, the history is much more recent. My Great-Grandfather Hidehiko Morioka arrived in the Hood River area in the late 1890s from Haga, Japan, in Okayama Prefecture. They cleared land for orchard production and farming and secured their own plots of land in Dee, Oregon. According to my grandfather, Harry Takeshi Morioka, it was a friend from Japan—Nobu Imai—who taught the families how to locate and pick matsutake mushrooms in the mountains of Oregon.

The attack on Pearl Harbor on December 7, 1941, dramatically changed the life of West Coast Japanese communities. With the subsequent passing of Executive Order 9066 by Franklin D. Roosevelt on February 19, 1942, my family—along with approximately 120,000 other West Coast people of Japanese descent—was forced to move into ten different concentration camps around the United States.

My family was sent to Tule Lake concentration camp in California and Minidoka concentration camp in Idaho, where they were kept behind barbed-wire fences. Most of the concentration camps were in arid, desolate areas full of dust and wind. They lost their freedom to wander through the woods. They lost their freedom to hunt matsutake, the very thing that defined and grounded them—the activity that connected them to the earth and to their community. My family members lost their properties, their livelihoods, their communities, their rights, their pride, and their honor.

After the end of World War II, my Japanese American family decided to return to the Columbia Gorge in Oregon. The families who returned were met with resistance and overt racism. They started picking up the pieces of their lives and rebuilding from scratch. As my grandparents' little Mazda truck rolled along the old roads through the wilderness, my grandfather would be playing Hawaiian music on cassette tape. My Grandmother Kazuko would be squeezed in the back cab, insisting that she was comfortable but always asking for the volume to be turned down. They would park the truck down the road from their starting point and backtrack through the woods, so as to not be followed (this was during the 1990s, when matsutake prices were at a high because of commercial demand). When other pickers asked if we had found any matsutake, we would reply, "Enough for soup." There were places in the woods we would not venture. As my grandmother would say, "That's where the Migaki family goes . . . we don't go down that way." There was understanding and respect out there in the woods.

My grandparents eventually started to share some of their stories with me in their later years. As the oldest grandchild, perhaps they knew that I would be the keeper of the family history. Perhaps I was just in the right place at the right time. Regardless, matsutake hunting is one of the traditions that has survived the generations. As an adult, I have found my own matsutake hunting grounds, and every fall, I return to them with my own family. My daughter, a Gosei, fifth-generation Japanese American, always seems to find the first matsutake of the season!

7. THE PRINCE

(Agaricus augustus)

A royal prize of a mushroom fit for a king, the prince (*Agaricus augustus*) is striking and hard to miss. Worthy of its noble name, it is tall in stature with a large, golden brown cap covered in scales. It is noted for its strong almond smell and flavor. This is a tasty mushroom and prized by mushroom hunters all over the world. The royalty in your castle—in fact, the whole kingdom—should experience *Agaricus augustus*.

The prince has chocolate brown spores at maturity.

Cap Characteristics

The cap is large in size (four to twelve inches). Usually blocky at first, it becomes convex to broadly convex or nearly flat when mature. The top of the cap is dry and whitish-colored, with a dense covering of brown to dark-brown fibrillose scales. It bruises yellow at the edges.

The cap of the prince is covered with brown fibrillose scales.

Underside of Cap

The cap underside has gills that are white to gray when young and turn chocolate brown to black when mature, never pink; they are free from the stem. A veil is present when it is young.

Stalk

The stalk is thick (greater than a half inch), tall (four to eight inches), and scaly with a large, skirtlike ring. The flesh is white and firm. There is no volva at the base.

Fragrance and Edibility

The prince has a sweet and almondy fragrance and taste. It is delicious and considered as one of the best edible mushrooms.

Habitat

Often gregarious, the prince can be found growing near conifers, roads, paths, gardens, parks, and grassy areas. Common on the West Coast, especially west of the Cascades and in the Coast Range, it can be found both summer and fall. A similar species, *Agaricus julius*, can be found in the Rocky Mountain Range.

Similar Mushrooms That Should Be Avoided

Toxic look-alikes include *Amanita* species, such as the death cap and destroying angel. These species have a volva at their base and white gills at maturity. They can emit a bad odor. They are described in more detail in this guide. Smith's amanita (*Amanita smithiana*) also has free gills and scaly stem with a ring, but the gills and spores are white, not dark chocolate brown like the prince, and it smells like old socks, not almondy. A few other similar species can be eliminated by noting the prince's white-gray gills turning to chocolate brown, never pink.

A PRINCE OF A STORY

Tim Giruadier

I had just moved to Port Townsend, Washington, and was renting a tiny A-frame perched atop a heavily forested glacial moraine. The first year I was there, I made a vegetable garden, which was no easy feat because the soil was largely comprised of rock from the moraine. After a month or so of growing vegetables, the largest mushroom button I'd ever seen (perhaps to this day) pushed up from the tilled soil. It was spectacular, and I enjoyed watching it grow every day.

It didn't take long for the landlord to come up and see what I was up to. When he saw the mushroom growing in the garden, he immediately recognized it and said, "That's one of the good ones. It tastes like almonds, you should eat it!" To myself, I thought, "I'm not going to eat that just because you say so." A few days later, a neighbor came over, and I showed him the mushroom and said, "It's supposed to taste like almonds." To which another friend said, "Yeah, cyanide smells like almonds," and he laughed at me.

By then I was curious, and I looked in the *Mushrooms Demystified* book to identify it. It was the prince *Agaricus augustus*. I sautéed it in butter and salt, and sure enough, it did taste like almonds. It was like it was sautéed in almond extract. I was truly amazed, and this was the mushroom encounter that got me hooked on learning about mushrooms and preparing them. Since then, I've occasionally found *Agaricus augustus*, primarily in the Oregon Coast Range, but none have compared to the size and flavor of this big beauty from Washington State.

Recipe:

ASPARAGUS
AGARICUS AUGUSTUS

The addition of the prince makes this easy stir-fry meal worthy of royalty. Serve it over rice. (And just for fun, try repeating "asparagus *Agaricus augustus*" as fast as you can ten times.)

INGREDIENTS:

1 bunch asparagus, chopped
1 large cap of the prince, chopped
4 broccoli flowerets
3 tbs. chopped green onion
2 tbs. olive oil
2 tbs. balsamic vinegar
2 tbs. butter

METHOD:

Fry mushroom in olive oil for 3 minutes at medium-high heat. Add asparagus and cover for 4 minutes, stirring occasionally. Add broccoli and balsamic vinegar, and cover for 3 minutes, stirring occasionally. Add butter and green onion, and cook for 3 minutes, stirring occasionally.

8. SHAGGY MANE

(Coprinus comatus)

Shaggy mane (*Coprinus comatus*) mushrooms are easy to identify—a popular edible and a great choice for the beginning forager. But there is a catch. They are perishable and liquefy in a couple days into an inky mess. So leave specimens that are beginning to degrade in the field. Only pick young, bright specimens, and plan to eat the shaggy manes within a day after you collect them. But wait, there is another catch. You should not drink alcohol when eating shaggy manes (there is an allergic reaction to mixing the two). I love shaggy manes with eggs in the morning. And remember—abstain from alcohol. Come on, you can do it for breakfast!

These distinctive mushrooms can be identified by their bright-white, cylindrical cap when they are young that turns gray with age. The caps of older specimens turn to ink, beginning from the bottom of the cap and moving up. There is no volva at the base (always check by carefully digging up the stalk).

The shaggy mane is also known as "shaggy ink cap," "lawyer's wig," and "shaggies."

Cap Characteristics

The cap is large and tall (four to ten inches) and shaped like a cylinder when young. It is very shaggy with large, light-brown scales on a bright-white cap. It transforms from a cylinder to a bell shape with age. The cap digests itself starting at the edges and moving up, turning the flesh into an inklike consistency.

Underside of Cap

The cap underside has a crowded gill layer free from the cap. White when young, they turn gray and then into an inky-black mess with age.

Stalk

The stalk is hollow, long (four to nine inches), and white. A veil is present, forming a loose ring on the lower part of the stalk. There is no volva at the base.

Fragrance and Edibility

Shaggy manes have a subtle earthly smell and flavor. They are delicious when freshly picked, but they won't keep for long. Eat them as quickly as possible, and toss them out before they turn into an inky slime in your refrigerator.

Shaggy manes decompose rapidly, so harvest only freshly intact cylindrical caps. These specimens are too mature to harvest.

I have included a recipe below for shaggy mane–crusted parmigiana. Shaggy manes are also good in soups and with risotto.

Many people claim to experience an allergic reaction from eating shaggy manes and drinking alcohol. It is prudent to avoid alcohol when eating shaggy manes.

Habitat

When conditions are right, you can find shaggy manes in a diversity of settings: forested areas, parking lots, lawns, compost piles, compacted soils, and other areas of disturbance. Shaggy manes' fruiting occurs in the late summer and early fall, often along roads and trails. They are gregarious just before a hard frost.

Similar Mushrooms That Should Be Avoided

The common inky cap *Coprinopsis atramentaria* is edible but causes an adverse reaction when combined with alcohol. It contains an amino acid that interferes with the metabolism of alcohol, causing alcohol toxicity. The common inky cap is differentiated from the shaggy mane by its lack of distinctive scales on the cap.

Fun Facts about Shaggy Manes

- Inky cap mushrooms, like the shaggy mane, have been used as ink for centuries. Numerous online recipes and YouTube videos demonstrate how to make ink from shaggy manes and other ink cap species.

- So what's up with the ink? Inky caps drop their spores by autodigestion. When spores on the crowded gills are ready to be released, the mushroom produces an enzyme that digests the flesh, causing the cap to turn to ink and curl upward. This exposes the gills to the wind, which disperses the spores. When it is all said and done, the whole cap becomes just a gob of inky goo.

Shaggy manes use autodigestion by turning to ink and thus exposing its spore mass to the wind.

- Even though shaggy manes are easily broken and delicate when handled, I have seen them pushing up through asphalt forest roads. It's perplexing—how can such a delicate mushroom break up an asphalt road surface? Simply astounding!

Recipe:

SHAGGY MANE-CRUSTED PARMIGIANA

INGREDIENTS:

shaggy mane mushrooms, sliced lengthwise
all-purpose flour for breading
eggs whisked with a splash of milk
grated fresh Parmigiano Reggiano
chopped fresh parsley
salt and pepper to taste
grapeseed or canola oil
lemon wedge for serving

METHOD:

Season the egg mixture with a pinch of salt and pepper, and combine with chopped parsley. Dip the shaggy manes into the flour, then into the parsley egg mixture, and sprinkle on Parmesan cheese.

In a large cast-iron skillet or nonstick surface big enough to hold the shaggy manes, heat oil until hot. Add the shaggy manes to the pan and cook for 4 minutes. Once the mushrooms are golden brown on one side, use a spatula to gently loosen them from the pan.

Flip the mushrooms and cook for 4 minutes on the other side or until the mushrooms are completely crisp and the crust is golden brown. Transfer the mushrooms to a wire rack or a paper towel to cool for a minute and remove excess oil. Serve immediately with lemon wedges or just by themselves.

9. PSILOCYBES

Gold caps, *Psilocybe cubensis*.

Psilocybin mushrooms, or *Psilocybes*, contain psychoactive substances and are commonly known as "magic" mushrooms. They have been used for millennia and continue to be used by numerous cultures in spiritual, religious, and recreational contexts. Well-documented ancient sites in Spain, North Africa, and Mesoamerica contain "magic mushroom" stones and motifs dating back thousands of years (Samorini 1992, Akers et al. 2011).

As revealed by the Forest Service website article "Teonanacatl Mushrooms: Flesh of the Gods," early Spanish explorers described the firsthand observations of ceremonial use of psilocybin mushrooms by the Aztecs in AD 1502. Interest in psilocybin mushrooms as a recreational drug increased during the 1970s and 1980s. This was inspired, in part, by the May 13, 1957, *Life* magazine article, "Seeking the Magic Mushroom," which featured R. Gordon Wasson relating his experience with the mushrooms in Mexico. In the 1970s, individuals such as Timothy Leary and Terrence McKenna promoted psilocybin use in the "hippie counterculture."

Though illegal for decades, research and general interest in psychedelics are "mushrooming" today. Numerous studies are underway using psilocybin for treating medical issues such as post-traumatic stress disorder (PTSD), addictions, depression, anxiety disorders, and cognitive decline. Laws are being considered to allow the regulated and medical use of psilocybin in some states. The popularity and availability of psilocybin mushrooms, both wild and cultivated, have made them widely harvested and consumed across the globe.

Paul Stamet's book *Psilocybin Mushrooms of the World* is a comprehensive guide to numerous species if readers are looking for an in-depth examination of the psychoactive mushrooms. Michael Pollan's book and Netflix documentary *How to Change Your Mind* is a relevant and deep dive into the body and mind effects of psilocybin mushrooms.

The four species of psilocybin mushrooms in *Fry, Thrive, or Die* are hallucinogenic. Other psychedelic mushrooms can also be found in the Western United States, such as *Psilocybe azurescens* (very potent) and some *Panaeolus* species. The experience of using psychedelic mushrooms is strongly dependent upon the *dose* (for example, a 0.3-gram microdose versus a 5-gram "heroic" trip), *set* (frame of mind of the user), and *setting* (the social and physical environment where the experience takes place). Laughter, enhancement of colors, lack of concentration, and muscular relaxation are common side effects. Heightened anxiety, leading to a "bad trip," can be the result of an unpleasant and unfamiliar setting. Higher doses can result in hallucinations and the inability to distinguish fantasy from reality. Many users find it preferable to ingest psychedelic mushrooms in a serene, natural environment and with friends who can provide support.

The biggest risk of consuming psychedelic mushrooms collected in the wild is confusing psilocybin mushrooms with the many varieties of little brown mushrooms that are poisonous and even lethal. Psilocybin mushrooms should not be confused with other poisonous smaller brown mushrooms with dark gills. The deadly

Galerina marginata is a "little brown mushroom" that resembles the *Psilocybe* group but has rusty-brown, not purple, spores. Poisonous *Psathyrella, Lepiota, Pholiotina,* and *Inocybe* species are somewhat like psilocybin mushrooms and must be avoided. For more intensive descriptions of little brown mushrooms, see David Arora's *Mushroom's Demystified*. Remember, when in doubt, throw it out.

GOLD CAPS

(Psilocybe cubensis)

Gold caps.

Gold caps (*Psilocybe cubensis*) are potent psychedelic mushrooms and a popular choice for cultivation by indoor gardeners. While there has been some recent relaxation in the penalties of possessing small amounts of psilocybin mushrooms, they are still illegal in most US states and in many countries.

Several strains of gold caps are cultivated indoors in containers.

Gold caps are also known as "golden teacher," "golden halos," "cubes," "bare head," "cubensis," and "magic mushroom."

Cap Characteristics

The cap is small to medium in size (generally a half inch to four inches in width). With a conic shape when young, it flattens as it matures. Smooth and sticky, the cap is brown and becomes pale to almost white at the margin, turning golden brown with age. When bruised, the cap turns blue. The cap can turn the same color as the gills because of falling spores from the gills of mushrooms located above.

Underside of Cap

The cap underside has gills that are never yellow brown or rusty brown. The gill attachment is narrowly attached to almost free attachment. They are cream-colored before turning dark gray to purple gray to purple black with maturity. The gill margins often remain whitish. The spore print is always an important diagnostic check and should be purple gray and purplish black. The veil under the cap is cobwebby when young.

Stalk

The stalk is generally medium in length (two to five inches) and medium in thickness. It is white and hollow and stains blue. A universal veil when young leaves a well-developed ring on the stalk. The veil can turn the same color as the gills because of falling spores underneath the cap.

Fragrance and Edibility

Gold caps have a light, sweet, woody smell and a taste reminiscent of baked pumpkin seeds. The concentration of psilocybin varies widely depending upon the strain, typically .5–2 percent of the dry weight of the mushroom. Doses vary depending upon the concentration of psilocybin in the strain. Typically, 0.2–0.5 dried grams would be considered a microdose. A usual "recreational" dose is considered 1–2 grams, while a strong to "heroic" dose ranges 2–5 grams.

Blue staining on a strain of *Psilocybe cubensis*. The concentration of psilocybin can vary by strain.

Habitat

Gold caps are common in Mexico and in tropical and subtropical areas. It occurs on dung, manure, and compost in its native habitat. Widely cultivated in the Western United States under artificial conditions, both indoors and outdoors, it is the most well-known *Psilocybe* species due to its ease of cultivation and wide distribution.

LIBERTY CAP

(Psilocybe semilanceata)

The liberty cap (*Psilocybe semilanceata*) is found widespread across the Northwest. While it has a lower psilocybin content than the gold caps, it still carries a mind-expanding punch. Though it grows naturally in the wild, the possession of this mushroom is illegal in most US states and many countries.

The liberty cap is also known as "magic mushroom."

Cap Characteristics

The cap ranges from tiny to small in size (generally less than one inch). It is shaped like a cone or a bell and has a knob or a nipple-like protrusion on the very top. The cap is brownish when moist, but the moisture fades to tan when dry. It has bluish and olive-colored stains and has radial lines at the edge.

Underside of Cap

The cap underside has gills narrowly attached (adnexed) to almost free. They are cream-colored gills before turning dark gray to purple, black, and brown with maturity. The spore print is always an important diagnostic check and should be dark purplish brown. The partial veil under the cap is cobwebby and quickly disappears with age.

Stalk

The stalk is slender (less than 1/8 inch), generally two to four inches long, and either slightly lighter or the same color as the cap. It usually stains blue, olive, or tan with handling.

Fragrance and Edibility

Liberty caps have a mild cheese and mushroom smell. They can be hallucinogenic, depending on the dose. The psilocybin content of liberty caps is lower than that of gold caps.

Habitat

You can find liberty caps in grassy areas. Saprophytic, it feeds off decaying grass roots. It occurs in moist temperate environments and is often found adjacent to areas containing sheep or cow dung, although not growing directly on the dung. It is widespread globally and occurs frequently in the Pacific Northwest west of the Cascade Mountain Range.

WAVY CAPS

(Psilocybe cyanescens)

Wavy caps (*Psilocybe cyanescens*) are potent psychedelic mushrooms native to the Northwest United States. This species has a high psilocybin content. And if you have this mushroom in your collection bag, you are breaking the law in most US states and countries around the world.

Wavy caps are also known as "wood chip psilocybe," "bluing psilocybe," "potent psilocybe," and "magic mushroom."

Cap Characteristics

The cap is small to small/medium in size (generally one to two and a half inches across). In coloring, the cap is caramel to reddish brown, darker when moist but fades to tan when dry. Dome-shaped when young, it turns flat and wavy as it matures.

Underside of Cap

The cap underside has cream-colored gills that turn purplish gray to purplish brown with maturity—never yellow brown or rusty brown. The gill attachment is narrowly attached (adnexed) to almost free. The spore print should be purplish gray and purplish brown. The partial veil under the cap is cobwebby and quickly disappears with age.

The underside of the wavy cap is cream-colored to purplish gray and purplish brown, never yellow brown or rusty brown.

Stalk

The stalk is slender (less than 1/8 inch), generally two to four inches long, and either slightly lighter or the same color as the cap. It usually stains blue, blue green, or tan with handling.

Fragrance and Edibility

Wavy caps have a mild mushroomy smell. It is a potent psychedelic mushroom.

Habitat

You can find wavy caps in wood chip areas, landscaping, mulched garden, and edges of wooded areas. It fruits gregariously in patches of hundreds to even thousands of mushrooms, generally in the fall when temperature first starts to drop significantly. Saprophytic, it feeds off decaying woody and mulched substrates. It is widespread in California and in the Pacific Northwest; it also occurs in Europe.

STUNTZ'S BLUE LEGS

(Psilocybe stuntzii)

Stuntz's blue legs (*Psilocybe stuntzii*) are psychedelic mushrooms that grow in conifer wood chips and bark mulch, woody debris, or new lawns in the Pacific Northwest. Discovered on the campus of the University of Washington by Dr. Daniel Stuntz, this psychedelic mushroom has a lower level of psilocybin than the other *Psilocybes* described in this section. Now, I will say this one last time—the possession of psilocybin mushrooms is illegal in most US states and many countries. If you decide to collect it for consumption, be careful because this *Psilocybe* closely resembles the highly toxic *Galerina marginata*. Spore prints are always advised. Stuntz's blue legs are also known as "blue ringers."

Cap Characteristics

The cap is tiny to small (usually half inch to one and a half inches) and generally two to four inches long. Its olive-green coloring fades to a pale olive brown or pale yellowish brown. It stains greenish blue when handled. The cap is darker when moist but fades when dry and is lighter toward the center. It is conic when young, expanding to umbonate or flat with maturity. The margin is striate when moist.

Underside of Cap

The cap underside has gills attached or narrowly attached (adnate or adnexed). Yellowish brown at first, they turn violet brown or chocolate brown to dark violet. The mature spore print is dark violet brown. It is important to take spore prints on Stuntz's blue legs so as not to confuse it with other toxic mushrooms. The mature gills are violet brown or chocolate brown to blackish violet.

Stalk

The stalk is slender (less than 1/8 inch thick) and stuffed with pith, becoming hollow with age. The base of the stalk is slightly enlarged. There is a fragile ring on the stalk that stains blue green. The stalk also stains blue green when handled.

Fragrance and Edibility

Stuntz's blue legs have a mild mushroomy smell. It is a psychedelic. The psilocybin content of Stuntz's blue legs is lower than that of gold caps or wavy caps.

Habitat

You can find Stuntz's blue legs in wood chip areas, landscaping, mulched gardens, new lawns, and the edges of wooded areas. This mushroom can fruit gregariously in patches of hundreds to even thousands of mushrooms, generally in the fall when temperature first starts to drop significantly. Stuntz's blue legs feed off decaying woody and mulched substrates. They are widespread in California and in the Pacific Northwest and can occur from August through November.

MUSHROOMS LESSONS LEARNED

Dr. Megan Frost

Dr. Megan Frost sneaks up on a large inky cap.

My love affair with mushrooms began only after I discovered my connection to the earth and the universe. Currently in my early forties, that connection feels so apparent, palpable, and ubiquitous. But growing up, I was indoctrinated with the belief that I was separate from this earth. I grew up in a place and time where it was a sign of prosperity to no longer need to grow your own food. With women finally accepted in the workplace and both of my parents working, processed food obtained under the fluorescent lights of a grocery store was the only food I knew. My introduction to mushrooms was as a pizza topping that my parents would order. I was indifferent toward them, and the idea that a person would walk into a forest, collect, and then eat mushrooms was fantastical. Additionally, I was raised in the Midwest, and the outdoor activities we flocked to revolved around sports. While that got us outside, it was onto fields of homogenous grass well groomed to standard heights. Nature was something to be cleared and controlled so that humans could live comfortably.

As I grew older and moved to Oregon, I was immersed into a place and culture that revered nature. I remember meeting my first Oregon friend who suggested hiking miles into a camping spot, and I wondered how I was going to carry my new five-person tent that far. I had not known there were places in the United States you could hike miles and then sleep. I also embarrassingly recall moving in with new roommates who wanted to plant a garden in the backyard—a foreign concept to me—and I had to watch them secretly out of the corner of my eye as they planted the vegetable starts. I did not know how to plant a garden. Over time, I began to feel at home in Oregon, connecting to the people and the world around me more than I ever had before. I also learned that button mushrooms sliced onto a pizza were not the only mushrooms out there.

As friends and farmers markets acquainted me with the vast array of mushrooms that could be consumed for enjoyment, my connection to the universe began to reveal itself to me. I realized I was not separate from the earth, but part of it. The mycorrhizal network epitomized the interconnectedness of the universe, allowing plants to communicate and transfer water and nutrients amongst one another. The decomposition of forest litter and wood by fungi unveiled how there is no death, just a transition to new life. These lessons led to a newfound understanding of my own physical, chemical, historical, and spiritual bonds to the universe, and this made me feel connected, yet insignificant. I awakened to the fact that the universe maintains a homeostasis regardless of what happens to my physical body. This brings me great comfort.

I had moved to Oregon to begin my surgical career. Over time that has evolved to focus on the surgical treatment of cancer. I learned the above lessons from mushrooms while simultaneously being exposed to death and suffering. Both my career and my understanding of the universe has advanced. I got to a point in my career where I was proud of how well I was able to treat a person's cancer based on our medical definitions, but I felt that I was inadequate in treating the person as a whole. I did not have the time nor opportunity to discuss at length how people feel emotionally and physically during or after treatments—what it was like to live with the fear of cancer always weighing on you, or how we could optimize one's death and make it as peaceful as possible.

I was ecstatic when I learned that Psychedelic Assisted Therapy (PAT), using psilocybin (which is a compound found in a mushroom), could be a tool to help my patients feel the same connection to the universe that mushrooms had taught me. That awareness of the relationship between us and the universe can help one cope with chronic pain, anxiety, and depression as well as alleviate fears surrounding cancer recurrence and death. I enrolled in a training program to learn more about PAT and how to integrate it into a person's health care. While we are just now on the cusp of PAT being legal, the lessons I have learned from this course and from mushrooms and nature are already allowing me to be more attuned to what my patients need to heal.

Like the mycorrhizal network links different plants, our bodies are coupled with plants through the cycle of oxygen and carbon dioxide. There are infinite links between our bodies and nature because we are simply a part of nature—not above it. And like a tree that dies in the forest can be decomposed by fungi to bring new life, when our physical bodies die, it is just a transition of our physical matter into the rest of the universe. That matter will be redistributed into innumerable new ways of life. Having this knowledge, I am better equipped to help my patients as I no longer view success only as stopping a cancer and preserving a life. I can be present with my patients and listen for what is important to them, focus on their quality of life, and help them to accept that death is not a failure or an ending but a part of our existence.

Every time I eat a mushroom, I taste the earth. Every time I see a mushroom growing, I am reminded of the interrelatedness of our universe. Mushrooms have been my teacher.

PSILOCYBIN MUSHROOMS— A REFLECTION

Dr. Pamela Kryskow, MD

Psilocybin mushrooms are a remarkable ally for humans in our journey. I have seen them heal people carrying the suffering burdens of life. In our work with palliative patients, they assist in the healing of old and new traumas so people can focus on living instead of dying. They assist us to become better versions of ourselves, shedding layers and triggers that no longer serve us. They allow us to become more thoughtful, kinder, more creative. The journey with psilocybin mushrooms allows us to be better humans and better coinhabitants of planet Earth.

Dr. Pam.

Dr. Kryskow is a Vancouver Island University adjunct professor, University of British Columbia clinical instructor, and Psychedelic Association of Canada founding board member.

THE STUTZII SECRET

David Steinfeld

These tree seedlings concealed thousands of tiny secrets.

It was in the late 1990s when I first learned of a mushroom called *Psilocybe stuntzii*. I was walking with my friend through a field of tree seedlings collecting mushrooms for his start-up business. Mike was looking for a specific group of mushrooms—those that form mycorrhizae on the roots of trees, important for reforestation and restoration projects. Since we were growing trees in our nursery, it was sometimes easier for him to come by and collect these mushrooms right out of our nursery beds instead of traveling into the forests to look for them. When Mike would come by, I would take a few minutes out of my busy schedule to look for mushrooms and catch up on life with him. As we strolled through the nursery beds, Mike would spot a mushroom deep between the trees and stop and pick it, then we would return to our lives.

In those days, we sowed conifer seeds in rows on the surface of the soil and covered them with a layer of sawdust. This practice protected the seeds while they germinated and created an environment for optimum germination. We had miles of sawdust-covered beds. Since mycorrhizal mushrooms had not yet formed on these beds, Mike would walk right by them. On one visit, however, Mike bent down and picked a very small mushroom that was growing out of the sawdust and began to scrutinize it. He put it down and picked another. And another. Then he looked up and said to me, "Dave, this little mushroom is psychedelic—*Psilocybe stuntzii* to be exact. They like to grow on sawdust." Then he looked across the field for as far as we could see and said, "Dave, you have a nursery full of magic mushrooms!"

I let this sink in for a minute and then realized that there were millions of these mushrooms in our nursery, enough to light up the city I was living in. I replied, "Mike, we cannot tell a soul about this, or we're screwed! Every hippy in the valley will be in here, including some of our employees, looking for these things! It will be a mess."

So . . . it became our secret.

I need to frame this moment in the culture of the time for you. We were going through a reefer madness period at work and in society, it seemed. It was a time when no one talked about smoking pot, let alone admitting we had ever tried it at some point in our lives. I had worked for years with fellow Forest Service employees, and marijuana was never brought up around the lunch table. It was too dangerous a topic—one that could hurt your reputation and possibly your job. I had given up smoking pot years before to become a family man, but on rare occasions, I would get high with friends. I remember sitting in a group of managers one day discussing whether we should impose random drug tests on our employees, which would include me. I knew I would likely flunk such a test because of the long half-life of THC. We were all a bit scared but dared not talk about it.

So explain to me then, why was I often seen out on my lunch breaks walking along the sawdust-covered beds? And why, when I returned to my office, did I have a pocket full of fungi? If you had met me at the time, you would have observed a harried middle manager, dealing with whatever was in front of him. But if you could have gotten into my head, you would have found a guy adrift in the doldrums of midlife. Every minute of my day was accounted for, and I was exhausted deep down. I hated being in management. I wasn't sure what life was all about. I just needed a breath of wind to fill my sails, and that was what the secret of this little mushroom brought me.

On the days I collected mushrooms, I would bring them home and lay the caps on a white piece of paper. And after a day, I would remove them and have a poster full of purple spore prints. I would post these sheets of paper on my office wall at home and imagine that I had captured the soul of these mushrooms. And while I played with the idea of marrying souls, I never tried. I had never taken psychedelics, and while I needed a little breeze in my life, I did not want the hurricane an unpredictable dose of mushrooms might deliver! There is little doubt I could have used a strong breeze at that point in my life, but I would have had to do it by myself. And I was too scared to do it alone. No, my relationship with *Psilocybe stuntzii* was different—it was one based on awe and respect and possibilities lost. But really, it was just the little secret in my coat pocket that brought a thrill to one man's simple life.

THE STONED APE HYPOTHESIS

The Stoned Ape Hypothesis is a theory pioneered in 1992 by Terence McKenna in his book *Food of the Gods*. McKenna postulated that psilocybin use caused the primitive *Homo erectus* brain to rapidly process and reorganize information that led to a rapid improvement in language, art, technology, and cognition. Basically, psilocybin enabled *Homo erectus* to evolve to *Homo sapiens*. Could it be that early humans were harvesting and eating magic mushrooms that grew out of the manure of the animals they were following? Did the higher consciousness induced by psilocybin also improve speech, community organization, and imagination that characterized human evolution? McKenna believed so.

Terrence McKenna died in 2000. His life was as interesting as the ideas he generated. He was an author, ethnobotanist, philosopher, and passionate advocate for psilocybin use. His journey was wide and twisting. He was a butterfly collector in Indonesia, a hashish smuggler in India, a magic mushroom grower in Northern California, and a shaman in Hawaii. His enthusiasm for psilocybin mushrooms began on a trip with his brother Dennis in the Columbian Amazon in the early 1970s. While he maintained his passion for the Stoned Ape Hypothesis throughout his life, his theory was rebuffed by the scientific community of his day as "extremely speculative."

But was McKenna "on to something" or just "on something"? New research into the mechanics of psilocybin and the human brain has led to a renaissance of McKenna's Stoned Ape Hypothesis. Studies are underway at New York University and Johns Hopkins University to see if psilocybin use can profoundly alter consciousness and induce physical changes in the brain. Other recent studies document a surge in activity in the primitive brain network associated with emotions, which is one of the reasons why psilocybin is currently being studied for patients suffering from anxiety, depression, addiction, post-traumatic stress disorder (PTSD), and existential distress following a terminal cancer diagnosis. Studies from Johns Hopkins University medicine indicate psychedelic treatment with psilocybin relieves major depression (Griffiths 2020) and existential anxiety associated with a terminal cancer diagnosis (Griffiths 2016). Studies underway at New York University, such as the Psilocybe Cancer Anxiety Study and a Double Blind Psilocybe treatment study for alcohol dependence, are just examples of continuing research on the subject.

Psilocybin enhances new connections in parts of the brain related to memory and emotions while reducing activity in the area calling the shots—the default mode network, or DMN. Researchers have described the DMN as the CEO of the brain, where chaos is contained and order is maintained. Without it, the brain would be off its leash, and the ego would be dissolved. This sounds terrible, and it can be terrifying, but perhaps there are times in a person's life to let the brain have a little freedom. Researchers at Yale (Kwan 2021) have found that new neural pathways and perceptions are stimulated in the brains of study participants on psilocybin, which might explain why they report being more connected to the world and less attached to their own bodies during these guided mushroom trips. Psychedelics seem to facilitate a level of mental flexibility that unlocks hardened habits associated with addiction and depression.

In the animal world, there are numerous examples of intoxication. Birds eat fermenting berries. Moths seek psychoactive flowers to drink the nectar. Humans have sought intoxicating beverages for hundreds of thousands of years. The Aztec and Mayan use of psilocybin mushrooms is well documented. Reindeer and shamans eat psychoactive *Amanita muscaria* in Siberia. There are indications that the Neolithic man, eight thousand years ago, consumed hallucinogenic mushrooms. In Africa, Algerian cave painting of Tassili include a depiction of a shaman with mushrooms sprouting all over his body. Tassili is the oldest known petroglyph depicting the use of psychoactive mushrooms (Soukapova 2011). In Eastern Spain, the Selva Pascuala rock shelter contains abstract drawings suggesting the use of hallucinogenic mushrooms a millennia ago. The five-thousand-year-old iceman Otzi, recovered from glacial ice in the Italian alps, was carrying three mushrooms on his body. Though the mushrooms were not hallucinogenic, one species was believed to be used as an anti-inflammatory or antibiotic and the other species as a fire starter.

But where is the evidence for the Stoned Ape Hypothesis? Perhaps such a theory can never be proven. Yet scientists have no consensus for the doubling to tripling of the human brain size between five hundred thousand and one hundred thousand years ago. Could it be that consumption of these mushrooms affected the human brain and was the catalyst that sparked early humans into imaginative behavior, community bonding, and spirituality? Many believe the Stoned Ape explanation is too simplistic and unsophisticated to have much merit. They may be right. But I have come to appreciate how little we know about fungi and how much there is left to learn. Is Terence right about our hunter and gatherer ancestors munching on magic mushrooms as they foraged their way across Africa? Could be. Did these small mushrooms trigger evolution of the human mind? I suppose it depends upon how far you are willing to speculate and how free you are willing to let your mind roam.

Terence McKenna

I met Terence at a mushroom conference tucked away in the Cascade Mountain Range a few years before he passed. He was an engaging, passionate, and sincere man who introduced himself to me as "shaman." Terence had a Cheshire cat smile and was obviously enjoying the conference presentations. This was a gathering of mushroom fanatics in a freewheeling celebration of the earth, soil, medicine, music, plants, decomposers, problem-solving, networked consciousness, and spiritual insight held on a Halloween weekend at the peak of mushroom season! The forests around the hot springs, where the conference was being held, contained a plethora of diverse mushrooms that we laid out on picnic tables for display at the end of each day. These included delicious edibles (like chanterelles, boletes, cauliflower mushrooms, and lion's mane) and some mind-altering *Psilocybes*, which found their way into a tea that brewed in a large cauldron on the back porch of the lodge. Let's just say the Halloween party that evening was festive and memorable.

Note—Paul Stamets, mycologist and psilocybin expert, explains the Stoned Ape Hypothesis in many YouTube videos.

10. OYSTER MUSHROOM

(Pleurotus ostreatus)

This edible mushroom can be found in the woods as well as in supermarkets. It is a workhorse species—used to clean up pollution, eaten for taste and health benefits, made into vegan leather, and molded into an alternative building material. What do you say about a fungus that is delicious to eat but also able to degrade cigarette butts and crude oil? That can be sautéed into an amazing stir-fry or grown into packaging material as an environmental replacement for Styrofoam? You'd likely say that it is a fantastic fungus!

The oyster mushroom (*Pleurotus ostreatus*) is a white-rot fungus that grows primarily on dead hardwood stumps and logs and fruits in a distinctive shelflike manner. The oyster mushroom (the name probably comes from the oyster shape of the cap) is the common name for a group of species or varieties (pearl, golden, king) with very similar characteristics, referred to collectively as *Pleurotus ostreatus*. For the typical mushroom hunter, it is not important to differentiate between these species and varieties since they all are edible. The oyster mushroom goes by a variety of names—including "pearl oyster mushroom," "tree mushroom," "tree oyster," and "hiratake."

Cap Characteristics

The cap is medium to large in size (generally three to six inches). The color is generally white, tan, gray, or brown in the wild. Some commercially produced oyster mushroom varieties have pink, green, or blue coloration. The range of color of this mushroom species is truly astonishing. The cap is also bald.

Commercial kits for oyster mushroom production include a pink variety.

Underside of Cap

The cap underside has gills that are white to cream to pale gray. There is no veil, ring, or volva. The gills are attached off-center to the stalk, and the spore print is white or pale gray.

Stalk

The stalk is short, thick, off-center. It is tapered and fused at the base. Sometimes the stalk is absent. It has a yellow-edged ring on the upper stalk.

Fragrance and Edibility

This mushroom has a slight seafood fragrance. The oyster mushroom is a tasty and popular edible. The mycelium grows aggressively in controlled sterile environments. Commercially produced, oyster mushrooms are available in many grocery outlets. Use it in any dish just like you would use other store-bought mushrooms.

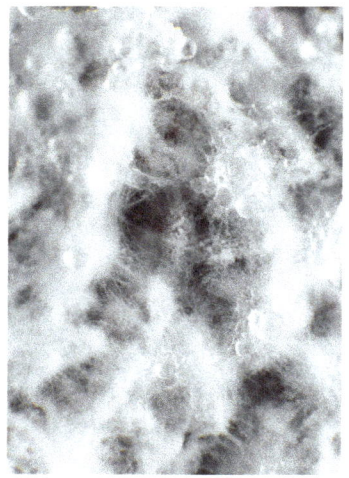

Oyster mushroom mycelium growing on rye grain.

Oyster mushroom growing indoors in a controlled humid environment.

Habitat

The oyster mushroom is very common and widespread in the fall in a variety of habitats. Fruiting is often abundant; you can find the oyster mushroom growing in dense shelflike clusters at the base of hardwoods stumps and dead hardwood trees. Commonly found on willow, oak, alder, and orchard trees throughout the Western United States, it fruits abundantly indoors as well in controlled humid environments.

Similar Mushrooms That Should Be Avoided

Toxic look-alikes include *Amanitas*—which, like *Pleurotus*, also have white gills. *Amanitas*, however, grow in soil (not on wood) and have a volva at their base and free gills. *Amanita* caps are not off-center from the stalk. Angel wings (*Pleurocybella porrigens*) are similar-looking to oyster mushrooms and are edible but fruit on the dead wood of conifers, not hardwoods. Angel wings are white, smaller, thinner, and flimsier than oyster mushrooms. Jack-o'-lantern mushrooms, *Omphalotus* species, grow in similar habitat and are off-centered from the stalk but have yellow to bright-orange caps.

VERSATILE FUNGUS

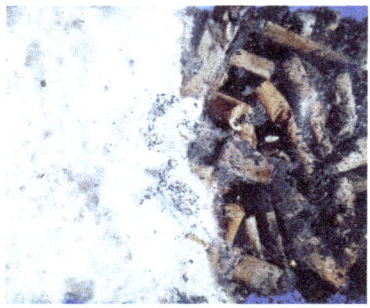

Oyster mushroom mycelium selected to degrade cigarette butts.

The oyster mushroom is not only delicious but also rich in fiber, antioxidants, nutrients, vitamins, and minerals. Recent studies indicate this mushroom is antitumor and immune-supportive. This is due to specific polysaccharides, known as beta-D-glucans. It also contains lovastatin, which is a type of statin that can lower cholesterol. Sixteen recent medicinal studies with oyster mushrooms are available at mushroomreferences.com.

The mycelium of this remarkable fungus breaks down toxic chemicals. Easy to grow, the oyster mycelium is being used to clean up pollution in "myco-restoration" projects. Oyster mushroom mycelium is voracious, eating its way through wood, paper, garbage waste, and even petroleum products (Sheldrake 2020). The mycelium excretes powerful enzymes and has an insatiable appetite to break down carbon bonds in toxic chemicals. The fungus can also absorb toxic heavy metals, such as mercury that may be present in toxic waste, so they can be disposed of safely.

Oyster mushroom mycelium is also being utilized in imaginative ways to produce structural materials. Mycelial networks like those produced by oyster mushrooms (and other rot fungi such as reishi and turkey tail) are extremely strong, fine, and interwoven. They produce materials from waste materials that are light and sturdy, resist water and decay, and can be formed into numerous structures. These structures are a natural mycelial alternative to Styrofoam, plastics, animal leather, and certain agriculture products (see ecovative.com). There is great interest in a variety of industries. For example, Adidas has a new proof-of-concept shoe—Stan Smith Mylo—that is made from mycelium-based Mylo material.

I've grown this mycelium in molds in my home and been amazed at how quick and easy it was to create lightweight objects. I've grown an assortment of mycelial objects, such as pots for plants, coasters, toys, and teddy bears for the grandkids.

A mycelial teddy bear grown in a mold using sterile wood waste.

Once you have the appropriate mycelial culture, all you need is some fine bark or wood shavings, a pressure cooker to sterilize the substrate, some flour and water to activate the culture, a sterile area to grow out the mycelial network, and an oven to dry the material. Hopefully, we will soon see a world that looks for alternatives to plastics. I believe mycelium from species such as reishi, turkey tail, and the amazing oyster mushroom is part of the solution. There are lots of interesting companies working with mycelium-based products on the web, such as ecovative.com, mylo-unleather.com, and buildwithrise.com.

Recipe:

COCONUT OYSTER MUSHROOMS WITH GINGER AND SCALLIONS

INGREDIENTS:

1 in. piece ginger root, chopped fine
3 garlic cloves, chopped fine
1 can (13.5 oz.) coconut milk, unsweetened
2 tsp. curry powder
1 tbs. soy sauce
12 oz. chopped oyster mushroom
1/2 cup mild chili peppers
2 scallions, chopped thin

METHOD:

Bring ginger, garlic, coconut milk, soy sauce, and curry powder to a simmer in a frying pan. Add oyster mushrooms, and stir and coat well. Simmer mixture, stirring occasionally until most of the coconut milk is evaporated—about 30 minutes. Add peppers and cook for 5–10 minutes. Season with salt to taste. Serve in a bowl and top with scallions.

11. DEADLY GALERINA

(Galerina marginata)

The problem with little brown mushrooms (LBMs) is that they all look pretty much alike. One of the most toxic LBMs is the deadly galerina (*Galerina marginata*), which contains the same amatoxins as the destroying angel and death cap *Amanitas*. It can make you very sick or even kill you if consumed. The deadly galerina has a "rap sheet" like most LBMs: cap is small and tan or brown, stem is slender, and it grows on wood and bark. The spore print is brown. It is very tough to distinguish from other LBMs!

It is also known as *Galerina autumnalis*, "funeral bell," "autumn skullcap," and "autumn galerina."

Cap Characteristics

The cap is small in size (usually about one inch). It has radial lines on the edge and is colored tan to brown or yellowish. The surface is bald and sticky when wet.

Underside of Cap

The cap underside has gills that are crowded and attached to the stem. The spore print is an important diagnostic check and is rusty brown or brown in color. The veil is present in young specimens.

Stalk

The stalk is generally one to four inches high and slender (less than 1/4 inch thick). A veil is present, first covering the gills then forming a small ring on the upper stalk that sometimes disappears with age.

Fragrance and Edibility

DO NOT CONSUME. DEADLY POISONOUS. The deadly galerina smells like flour (farinaceous). It contains the same amatoxins as two *Amanitas* described in this guide (the destroying angel and the death cap). In general terms, avoid eating little brown mushrooms (LBMs). The edibility of most little brown mushrooms is unknown, and many may be toxic.

Habitat

The deadly galerina is a saprobe that grows on wood or buried wood. Be aware—if the wood is buried, it might appear as if it is growing in soil. Common and widespread throughout the Western United States and the Northern Hemisphere, the deadly galerina has a long fruiting season and can fruit several times from mycelium in the same log or stump throughout the year.

Similar Mushrooms That Should Be Avoided

The deadly *Galerina marginata* resembles the psychedelic liberty cap (*Psilocybe semilanceata*) and several other *Psilocybes*, such as *Psilocybe stuntzii*. Caution, experience, and expertise are necessary to avoid consuming the deadly galerina along with *Psilocybes*. I have seen the deadly galerina growing on wood chips inches away from *Psilocybe stuntzii*. Unlike the *Psilocybes*, which have purple-brown spores, the deadly galerina has rusty-brown or brown spores. So if you are out collecting, always carefully inspect every mushroom in your basket you intend to eat. It is a good idea to conduct spore prints on mushrooms that are questionable. Poisonous *Psathyrella* and *Inocybe* species are also somewhat similar to the deadly galerina. For a more detailed description of little brown mushrooms, see David Arora's *Mushroom's Demystified*. When in doubt, throw it out.

12. TEETH FUNGI

Easy, nutritious, and delicious, the teeth fungi are easy to identify, important for your health, and downright fantastic edibles. The downward-pointing spines, or "teeth," define this otherworldly group of forest fungi. The four teeth fungi I describe in this section are fun to find and quite tasty. The hedgehog looks much like a chanterelle in color and texture, but it has spines instead of folds beneath the cap. The bear's head and the comb hericium form a branched array of magnificent icicles, while the lion's mane is a dense mass of long, majestic spines originating from a single base. Many teeth fungi, such as lion's mane, are being investigated for their health and medicinal qualities.

HEDGEHOG

(Hydnum repandum)

The hedgehog's (*Hydnum repandum*) white to pale-orange cap and spines on its underside makes the hedgehog easy to identify. It has no look-alike poisonous fungi, so it is a great choice for the beginning forager.

The hedgehog is also known as "sweet tooth," "wood hedgehog," and "belly button hedgehog" (*Hydnum umbilicatum*).

Cap Characteristics

The cap is small to medium in size (two to six inches). It is white to pale orange to orange—very similar in color to the golden chanterelle. It has a dry surface that sometimes develops scales with age. A related species, the "belly button hedgehog" (*Hydnum umbilicatum*) has a "navel" hole in the middle of the cap and is smaller and thinner than *Hydnum repandum*. It is equally good to eat.

Underside of Cap

The cap underside is white to pale orange in color, with spines, or "teeth." Distinctive!

Stalk

The stalk is generally two to four inches long and white to pale orange. It is smooth, has no veil, and brittle, not stringy or leathery.

Fragrance and Edibility

The hedgehog has a fruity fragrance and sweet and nutty flavor. It is delicious and one of the best edible mushrooms. The underneath spines can accumulate litter and soil, so clean it with a brush before placing it in your bag or basket.

Habitat

Often clustered, hedgehogs can be found in abundance. The species prefer coastal and mild climates in winter and early spring. It is widely distributed in conifer forests and occasionally occurs near oaks. A mycorrhizal forest species, it is often found in association with yellow foot chanterelles.

Some Similar Mushrooms That Should Be Avoided

None!

LION'S MANE

(Hericium erinaceus)

Lion's mane can get quite large, and it fruits on the same log or stump year after year.

The lion's mane, with its ghostlike spines, is simply breathtaking to behold. Do you see a ghost or the old man's beard? It is another great choice for beginning foragers because of its distinctive appearance and the fact that it has no look-alike poisonous fungi to confuse it with. Sister fungi to the lion's mane are bear's head and comb hericium.

The lion's mane is also known as "old man's beard," "bearded tooth," and "pom pom du blanc."

Form and Size

Lion's mane has a single cluster of spines hanging from a single dense base. Medium to large, it often reaches several pounds in weight. The spines are one to three inches long at maturity.

Color

The color is white flesh and spines. It can develop a yellowish to orange-brown tinge with age.

Fragrance and Edibility

The lion's mane has a faint fish odor similar to shellfish. It has a crab or seafood taste and texture and is delicious.

Habitat

Lion's mane occurs in Northern California, Oregon, and Washington on oaks and other hardwoods. Also found in other parts of North America, Russia, and Europe, it occurs on living hardwoods in wounds, dead standing hardwoods, hardwood logs, and stumps. It fruits in the same dead standing or dying hardwood, log, or stump year after year. Lion's mane is now commercially cultivated and available in many growers' markets, restaurants, and supermarkets. One of the best edible mushrooms, it can be marinated, sautéed, barbecued, stir-fried, or baked. Add clean and sliced lion's mane to meat, fish, pasta, or seafood substitute dishes. It is a well-documented medicinal mushroom that can also be dried and made into powder capsules, tea, or tincture.

Lion's mane is also commercially cultivated and becoming more common in markets and restaurants.

Or collect your own.

Some Similar Mushrooms That Should Be Avoided:

None!

Lion's Mane Medicinal Properties

There has been an explosion of research and health products surrounding the lion's mane mushroom and its extracts. The bottom line—it's good for your brain. Recent research has shown that lion's mane has a variety of health benefits and contains seventy beneficial metabolites. Much of these benefits surround improving human brain function and treating or preventing human neurological and cognitive decline (Lai et al. 2013, Obara et al. 2008). If you are one of the millions eating fresh lion's mane or the supplements made from it, you already know that!

Scientists have found that lion's mane may protect against dementia, reduce mild symptoms of anxiety and depression, and help repair nerve damage (Khan et al. 2013). The mushroom-references.com website contains the abstracts of thirty-seven recently published scientific studies on lion's mane. Many of these studies focus on neuropathways, cognitive function, depression, and anxiety. A variety of compounds present in lion's mane, such as water-soluble hericenones and erinaçines, have been found to induce synthesis of nerve growth factor, which the brain uses to maintain neuron sensory performance (Phan et al. 2014).

While lion's mane studies point to improvement of memory and concentration, lion's mane also has strong immune-boosting, anti-inflammatory, and antioxidant properties (Chong et al. 2019). It has been shown to lower the risk of heart disease, cancer, ulcers, and diabetes in animals. While the current research is promising, more human studies are needed to develop practical health applications for lion's mane and its extracts.

There has been an expansion in the availability of lion's mane products in recent years. Capsules, powders, extracts, and fresh fruiting bodies grown on grain or sawdust are widely accessible in stores. You have many options to find the delivery method that best meets your dietary and lifestyle needs.

Paul Stamets with lion's mane and the author with hen of the woods.

Lion's mane contains constituents that can be nutrients for your neurons and immune system.

—Paul Stamets

BEAR'S HEAD

(Hericium abietis)

A welcomed addition to the menu, the bear's head grows in the same spot every year and can be harvested annually on your "happy bear's head day." Bear's head is another great choice for beginning foragers because it is easy to identify and has no look-alike poisonous fungi. It is distinguished from lion's mane by its multibranched clusters of spines hanging from multiple branches. It is similar to its sister mushroom comb hericium but is larger, denser, and grows on conifers, not hardwoods.

Bear's head is also known as "goat's beard" and "conifer coral."

Form and Size

The bear's head does not have a cap but has numerous branched clusters of spines hanging from cascades of branches. Medium to very large, it can often reach ten pounds and sometimes as large as thirty to forty pounds. The spines are 1/4 to 1/2 inch long.

Color

The color is white for the flesh and spines. It can develop a yellowish to yellowish-orange tinge with age.

The bear's head is a welcome addition to the kitchen. Mature specimens develop a yellowish tinge but are still good to eat.

Fragrance and Edibility

The bear's head has a mild fish odor. It is delicious and similar to fish in taste and texture. Add clean and sliced bear's head to meat, fish, pasta, and soup dishes. It can be marinated, sautéed, barbecued, stir-fried, or baked.

Habitat

Common in Northern California and the Pacific Northwest, the bear's head is a saprobe that lives on dead standing conifers, conifer logs, and stumps. It fruits in the same place year after year.

Similar Mushrooms That Should Be Avoided

None! Bon appétit.

COMB HERICIUM

(Hericium coralloides)

Like a beautiful Christmas tree covered in a December snowstorm, comb hericium is more open and delicate than its sister fungi, bear's head and lion's mane. It is another great choice for beginning foragers and has no look-alike poisonous fungi.

The comb hericium is also known as "coral hedgehog," *Hericium ramosum*, and *Hericium americanum*.

Form and Size

The comb hericium does not have a cap. Instead, it has numerous long-branched clusters of delicate spines hanging from an open framework of multiple branches. It is generally smaller than the bear's head, and the spines occur in rows along an open, branching network. The spines are 1/2 to 1 1/2 inches long.

Color

The flesh and spines are white and can develop a yellowish to yellowish cream or buff with age.

Fragrance and Edibility

The comb hericium has a faint fish or crab smell. It is delicious and more delicate than lion's mane and bear's head, but not as fleshy. It should be slowly cooked.

Habitat

Common in the Western United States, this saprobe occurs on dead standing hardwoods, hardwood logs, and stumps. It fruits in the same place year after year.

Some Similar Mushrooms That Should Be Avoided

None!

Recipe:

SAUTÉED HEDGEHOG

INGREDIENTS:

1 lb. hedgehog mushrooms
1 tbs. butter
1 tbs. olive oil
1 clove garlic (optional)
1 tsp. fresh thyme
1 tbs. chives, minced
salt to taste

METHOD:

Place clean, sliced hedgehogs in a frying pan with butter or oil and sauté for 5 minutes. As simple as that! Add a pinch of salt, thyme, and chives, and feast on a delicious combination of fruity and nutty flavors.

Recipe:

LION'S MANE "CRAB" CAKES

INGREDIENTS:

2 lbs. fresh lion's mane mushroom
1 egg
1 cup bread crumbs
1 medium sweet onion, diced
1/4 cup mayonnaise
1 tbs. Worcestershire sauce
1 tsp. Old Bay seasoning
1 tbs. Dijon mustard
salt and pepper to taste
2 tbs. olive oil
fresh lemon (optional)

METHOD:

Shred lion's mane into small pieces similar to the size of flaky crab. In a large bowl, combine egg, mayonnaise, onion, Worcestershire sauce, Old Bay, Dijon mustard, salt, and pepper. Mix in lion's mane and bread crumbs and form into equal-sized patties about 1/2 to 3/4 inch thick. Sauté in olive oil over medium heat for 2–3 minutes on each side until golden brown. Add a squeeze of lemon (optional).

13. BOLETES

The king bolete, *Boletus edulis*.

The edible boletes are scrumptious . . . enough to make Italian Americans dress up in suit jackets and polished shoes to forage in the forest (see "My First Mushroom Hunt: Porcini" section). While some boletes are poisonous, they are distinct and avoidable. *Fry, Thrive, or Die* can help keep you from dangerous lookalikes.

The boletes have a fleshy cap and central stalk, similar to many mushrooms but are distinctive beneath the cap by having a spongy layer of pores instead of gills. The pores are small tubes that contain the spore, or "seeds," of the bolete mushrooms. Boletes are mycorrhizal with trees, so they are found in forest environments and quite common across the Western United States. There are many species of boletes. I have included six common species that include three edible and three poisonous mushrooms; they are all equally distinctive species.

To identify the wider array of bolete species, refer to *Mushrooms Demystified* by David Arora, which covers the group in over fifty pages of comprehensive detail. Identifying this wide array of bolete species will take years of study and experience. Suffice it to say, learn to identify the porcini, the admirable bolete, and the manzanita bolete. They are delicious! Avoid boletes with red pores, a red stalk, or those that bruise blue.

PORCINI

(Boletus edulis)

The porcini does not have red pores or red flesh. It does *not* bruise blue under the cap or on the stem. The stalk is never bright yellow or red. It is one of the world's most prized edible mushrooms, and Italians are crazy about this mushroom.

Porcini is also known as "king bolete," "cep," "steinpilz," and "penny bun." The spring version in the Cascade Mountain Range is known as "spring king," or *Boletus rex-veris*.

Cap Characteristics

The cap is medium to large in size (generally three to nine inches across). It is colored yellow brown to reddish brown. It can be white when young beneath the forest duff layer, and it has smooth texture. A large porcini can weigh a couple pounds. One mushroom can be a meal!

Underside of Cap

The cap underside has a sponge layer. White and minute when young, the pores turn yellow brown to olive brown with age. The pores also do not bruise blue.

Stalk

The stalk is three to six inches long generally and an inch or more in thickness. It is often somewhat bulbous at the base, especially when young. The stalk is white or brown, never yellow or red. The top is finely netted. It has no veil. The base of the stalk is not bright red or rhubarb in color. The flesh is never red or turns blue when cut or bruised.

Fragrance and Edibility

The porcini has a mild, pleasant mushroomy fragrance. It is delicious fresh or dried and rehydrated, and it has a very nutty flavor. If the sponge layer under the cap is old or buggy, cut them out and toss them out. If the porcini is young and solid and the pores are whitish or yellow brown (and are in good shape), cook them with the mushroom or cut them out and use them fresh or dried in risotto, pastries, soups, and sauces. For later use, porcini can be cut into quarter-inch thick slabs and dried in a dehydrator. Store in an airtight container. Rehydrate in a pan of warm water for thirty minutes then cook. The porcini taste is concentrated by drying and rehydrating.

I love porcini cut in quarter-inch slabs, basted in olive oil, and baked for fifteen minutes then topped with good cheese and prosciutto. Another delicious way to prepare porcini is to dip quarter-inch slices in egg batter and bread crumbs, and fry in olive oil for five minutes. There are more recipes at the end of the bolete section.

Habitat

The porcini is mycorrhizal, so it occurs in forest areas and the margins of the tree root systems. Conifers such as fir, pines, and spruce in higher elevation forest are prime habitats, but porcini also occurs sometimes in lower elevations in oak, birch, and madrone habitat. Fruiting occurs in the fall about four weeks after the first drenching rains. In the Western United States, there is a May/June higher-elevation fruiting of a porcini called the "spring king bolete" (*Boletus rex-veris*), and it looks identical to the fall porcini but is now considered a different species than *Boletus edulis*.

The spring king (*Boletus rex-veris*) looks identical to the fall porcini and is also delicious. It also gets less buggy than the fall porcini. Spring king taste and preparation are similar to the fall porcini.

Similar Mushrooms That Should Be Avoided:

Avoid the bitter bolete (the cap underside turns blue when bruised, and the stalk is partly or completely red or rhubarb in color). The red-pored bolete and Satan's bolete are also poisonous, and they stain blue.

SATAN'S BOLETE

(Rubroboletus satanas)

The Satan's bolete is conspicuous. It has a bright-blue staining when bruised under the cap and on the stem. It has red pores when mature and a very bulbous base.

The Satan's bolete is also known as "devil's bolete" and *Boletus satanas*.

Cap Characteristics

The cap is medium to large in size (generally three to ten inches across). It is dull grayish and olive gray in color and has velvety texture. This cap bruises blue quickly when squeezed or cut. The specimens are sometimes very large, and individuals can weigh several pounds.

Underside of Cap

The cap underside has a sponge layer. The pores are yellow to red and stain blue when bruised.

Stalk

The stalk is three to six inches long generally and an inch and a half or more in diameter. It is very bulbous at the base. The top of the stalk is finely netted. It has no veil and is brown to bright red in color, turning blue quickly when cut or bruised.

Fragrance and Edibility

AVOID CONSUMPTION. POISONOUS. This Satan's bolete has a putrid smell when mature and can cause severe gastric distress if eaten.

You are hunting porcini, and you see a large bump in the forest floor, your heart soars . . . then comes crashing to earth or below when you discover it is Satan's bolete.

Habitat

Satan's bolete is mycorrhizal, so it occurs in forest areas and the margins of the tree root systems. Conifers such as fir, pines, and spruce in higher-elevation forest are prime habitats. Fruiting occurs in the fall about two weeks after the first drenching rains. Satan's bolete also occasionally occurs in lower elevations in oak and madrone habitat.

Similar Mushrooms That Should Be Avoided

Avoid the red-pored bolete and the bitter bolete.

RED-PORED BOLETE

(Rubroboletus pulcherrimus)

The red-pored bolete has orange to red pores and dark-red to purplish stalk. It rapidly stains blue to black when bruised under the cap and on the stem. The base is without a bulb.

It is also known as "slender red-pored bolete" and *Boletus erthropus*.

Cap Characteristics

The cap is medium to large in size (generally three to eight inches across, occasionally as large as ten inches). It is reddish to purplish brown in color and has lots of fiber and scales on top. This cap bruises blue quickly when handled or cut.

Underside of Cap

The cap underside has a sponge layer. The pores are orange to red and stain blue when bruised.

Stalk

The stalk is three to six inches long generally. The stalk is thick, an inch or more. It is not bulbous at the base. The top is finely netted red. There is no veil, and it is orange to red brown in color, turning blue or blue black quickly when cut or bruised.

Fragrance and Edibility

AVOID CONSUMPTION. POISONOUS. The red-pored bolete has a mild smell and can cause severe gastric distress if eaten. It is the only bolete that has been documented to kill someone.

Habitat

The red-pored bolete is mycorrhizal, so it occurs in forest areas and the margins of the tree root systems. Conifer forests are prime habitats. Fruiting occurs in the fall. Also known to occur in association with tanoak, it is widely distributed across the Western United States.

Similar Mushrooms That Should Be Avoided

Avoid the Satan's bolete. The blue or blackish staining, slenderer stem, and lack of bulbous base distinguish the red-pored bolete from Satan's bolete. The bitter bolete is similar and also causes gastric distress. The bitter bolete has a tan or buff cap that is cracked on top.

BITTER BOLETE

(Caloboletus rubripes)

The bitter bolete has a red stalk, without netting, and has a tan or buff cap that develops cracks with age. The red stalk and the blue staining of the bitter bolete will keep you from mistaking it for the porcini.

Other common names include "red-striped bolete" and "red-stemmed bitter bolete."

Cap Characteristics

The cap is medium to large (generally three to eight inches across, occasionally as large as ten inches). It is tan or buff in color and not sticky or slimy. It cracks with age and bruises blue when cut.

Underside of Cap

The cap underside has a sponge layer. The pores are yellow and stain blue quickly when bruised.

Stalk

The stalk is three to six inches long generally and an inch or more in thickness. It is thicker but not bulbous at the base. The entire stalk is red or rhubarb in color. The top is not finely netted. There is no veil. The flesh is whitish to pale yellow, turning blue quickly when cut or bruised.

Fragrance and Edibility

The bitter bolete smells unpleasant. Cooking does not improve the smell or taste. Very bitter, it is not deadly poisonous but causes gastric distress to some people.

Habitat

The bitter bolete is mycorrhizal, so it occurs in forest areas and the margins of the tree root systems. Conifer forests are prime habitats, but it can also occur in oaks. Fruiting occurs in the fall. It is widely distributed across the Western United States.

Similar Mushrooms That Should Be Avoided

Avoid the red-pored bolete and the Satan's bolete.

MANZANITA BOLETE

(Leccinum manzanitae)

The manzanita bolete does not stain blue when bruised, does not have red pores or a red stem, and has pronounced brown or black dots (scales) on the stems. The cap is sticky or slimy when wet, and it occurs commonly in the fall under manzanita or madrone.

The manzanita bolete is also known as "madrone bolete."

Cap Characteristics

The cap is generally medium in size (three to six inches). It is sticky or slimy when wet and often has leaves or needles stuck to the cap.

Underside of Cap

The cap underside has a sponge layer. The pores are white or grayish to olive and do not stain blue when bruised.

Stalk

The stalk is generally medium to long (three to eight inches) and greater than half an inch thick. It is white with pronounced brown and black dots (scales). There is no veil. The flesh is white. Mature specimens sometimes darken to a purplish gray when handled.

Fragrance and Edibility

The manzanita bolete is edible. It is not as nutty-flavored as the porcini, but the preparation and storage are similar to the porcini.

Habitat

The manzanita bolete is mycorrhizal and occurs in specific association with manzanita and madrone root systems primarily in California and Oregon. Fruiting occurs in the fall. A similar aspen bolete (*Leccinum insigne*) and birch bolete (*Leccinum testaceoscabrum*) occur under aspen and birch trees respectively and have the pronounced brown-and-black dots on the stem like other *Leccinum* species. The aspen and birch boletes are also edible.

Similar Mushrooms That Should Be Avoided

Avoid the bitter bolete, red-pored bolete, and Satan's bolete. *Leccinum manzanitae* does not have red pores or red flesh and does not stain blue.

ADMIRABLE BOLETE

(Aureoboletus mirabilis)

The admirable bolete does not stain blue when bruised. It has pronounced velvet top and greenish-yellow pores. The cap is not sticky or slimy when wet. It is the only bolete that occurs on logs or buried wood.

The admirable bolete is also known as "velvet-topped bolete" and "bragger's bolete."

Cap Characteristics

The cap is medium in size, generally two to seven inches across. Its color is maroon brown or red brown. The cap not sticky or slimy when wet and has a soft, velvety top.

Underside of Cap

The cap underside has a sponge layer. The pores are yellow and greenish yellow and turn dark yellow when bruised; they do not stain blue when bruised.

Stalk

The stalk is generally medium to long (three to seven inches) and greater than half an inch thick. The stalk has a similar color as the cap, maroon brown or red brown, with a conspicuous fine netting. There is no veil. The flesh is dull white to yellow brown.

Fragrance and Edibility

The admirable bolete has a pleasant and mild lemony smell. It is a good edible and has a lemony flavor. Preparation and storage are like that of the porcini.

Habitat

The admirable bolete fruits in logs and buried wood. The wood habitat indicates it could be saprophytic, but some reports indicate this bolete is mycorrhizal with hemlock whose roots also associate with rotting wood. It is widespread in California conifer forests and throughout the Pacific Northwest.

Similar Mushrooms That Should Be Avoided

Avoid the red-pored bolete, bitter bolete, and Satan's bolete. The velvety top, bright yellowish or greenish pores, and growth on wood location make the admirable bolete distinctive.

MY FIRST MUSHROOM HUNT: PORCINI

My first mushroom hunt was in the fall of 1964. I was eight years old. I got to spend considerable time with my Grandpa Ernesto in East Los Angeles. Grandpa was an Italian immigrant from the little town of Caluso in the Piedmont area of Northern Italy. Like a lot of Italians of his generation, he was a lover of mushrooms. He had spent many hours collecting mushrooms in the "old country." In fact, he would tell me how he worked for a family that had trees that produced truffles. I would imagine trees filled with chocolate truffles but later learned they were Italian white truffles. Ernesto would haul soil from trees that were known producers of white truffles and transplant the soil near trees that were not producing truffles in hopes the soil transfer would induce fruiting.

Grandpa's friends in his Los Angeles neighborhood were also Italian immigrants, and they spent time playing boccie, smoking cigars, and talking how good the food was back in Italy. One day I heard them talking about going to hunt "little pigs." I was disgusted at the thought they were really going to hunt baby pigs.

Turns out, "little pigs"—or *porcini* (*Boletus edulis*) in Italian—literally means "piglets." It is a term that was first used by the ancient Romans to describe porcini.

That autumn, the porcini were "popping" in the San Bernardino mountain area of Southern California, and it was time to go on the hunt. As Ernesto's grandson, I was welcome to come. My grandpa dressed in his finest clothing for the hunt. He donned his dress coat, fedora hat with feather, and his polished Italian shoes. The car pulled up bright and early with three of my grandpa's Italian friends dressed in their "finest" for the hunt. They looked like they were going to a wedding or a funeral, which was an indication of how important this event was in their lives.

The windy road and the cigar smoke were nauseating, and I was happy for each stop to get out of the old car and breathe fresh air and explore the woods. Every stop, I found a few tiny brown mushrooms of little interest. But the "guys" were really encouraging me to find porcini with every step. Early afternoon, with my grandpa's help, I stumbled across a large bump cracking the soil surface. He handed me his forked metal rod that he also used to pull weeds in the garden. I popped the porcini out of the ground, and it lay on its side. I could see a plump, white stem and a big, reddish-brown cap. It seemed huge to me. To see the joy in my grandfather's eyes and the merriment of his friends was a memory I will never forget. Four old Italian men, in the mountains of America, celebrating a young boy's coming-of-age discovery. They made me feel like I had won the lottery. I was on my first treasure hunt in the forest!

When we got home, Grandpa let me slice up the porcini in thin steaks. I had never used his big, sharp knife before, and he showed me how not to cut off my fingers. I dipped each porcini slice in egg and bread crumbs, and Grandpa fried them in olive oil. I can't say it was my favorite-tasting meal at eight years old, but the joy of eating something that I had found in the woods was unforgettable. Grandpa devoured his slices. I had brought home dinner! I remember Grandpa gave me a big bowl of spumoni ice cream after dinner. I felt like a "king."

PORCINI CULINARY USES

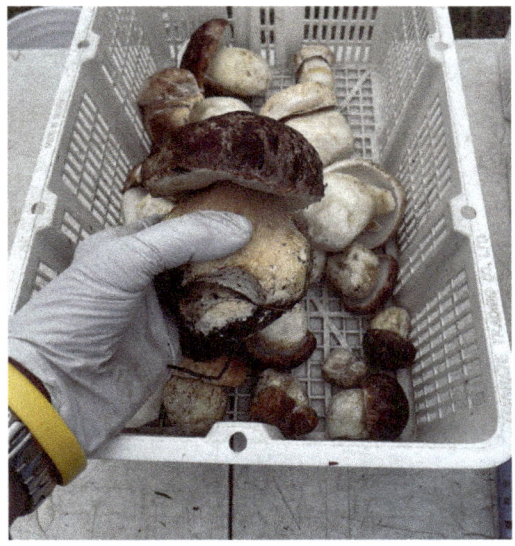

These young king boletes are just right for the kitchen.

Porcini is a tasty addition to all your favorite Italian dishes:

- In spaghetti sauce
- In sauces and soups
- Fried after dipped in egg and bread crumps
- As a topping for chicken, steak, or fish
- In any pasta recipe, especially risotto and gnocchi
- Caps coated in olive oil and grilled
- Chopped fine and cooked to a paste to serve on bread or with bruschetta
- A topping for pizza

Recipe:

PORCINI MUSHROOM WITH GRILLED CHEESE

Chef James Daw

INGREDIENTS:

porcini mushrooms, roughly 3 oz.
slices of good sourdough bread
3 tbs. of basil mayonnaise
4 slices of Tillamook white cheddar cheese
4 slices of prosciutto
2 tbs. butter

METHOD:

Preheat oven to 325°F. Slice porcini into quarter-inch slices and rub with olive oil. Place on sheet pan and cook for 15 minutes. Assemble the sandwich: bread, mayo, cheese, prosciutto, porcini. Butter the outside of the bread, and cook over medium heat until golden brown.

Recipe:

SIMPLE PORCINI RECIPE

Here's a *simple* recipe that really makes the porcini flavors pop. If you use dried porcini, rehydrate it for 30 minutes in warm water then drain.

INGREDIENTS:

4 cups porcini mushrooms
2 cloves of garlic
3 tbs. olive oil
4 cups ripe tomatoes
2 tbs. Italian herbs like basil, oregano, thyme

METHOD:

Warm the olive oil over medium heat. Mince the garlic, and sauté for about 3 minutes with your desired herbs. Chop the porcini mushrooms, and add to the garlic and herbs. Cook for about 5 minutes or until it looks like the mushrooms have released all their water. Add the chopped tomato and their juice, reduce the heat to low, and simmer for 20 minutes. Add some white wine if the mix gets too dry. Add to any meal and/or as an appetizer with bread.

14. MOREL AND FALSE MORELS

Morels: What, Where, and When?

Morels are diverse, delicious, and stealthy. They are magic in your mouth. The joy of finding one in the forest or field is equal to the incredible dishes you can dream up in the kitchen. The various species camouflage themselves, but the more you hunt, the easier it is to find these treasures. I love being a hunter and gatherer, especially when it comes to morels. I love debating the types of morels, where I think they will occur, and when the harvest season will arrive. I do this with my friends, family, or for that matter, anyone who will listen. Morels are treasured by gourmet chefs and relentlessly hunted by morel maniacs. They are perhaps the most striking, highly prized, and widely sought after of all the aboveground-fruiting fungi. Have you ever bought morels? You probably choked when you saw the price. But you know what, they are worth every penny!

Morel hunting is an artform that is handed down from generation to generation.

They can be prolific fruiters but sometimes difficult to find since they blend in so well with their environment. You can be staring at a piece of ground and not see them—then suddenly, like magic, you see one. Then you walk around a bit, and you see another one, and another one, and then it dawns on you that you had been standing in a large patch of morels for minutes before seeing a single one. It's humbling.

What Are the Types?

The names of various morel species have been surrounded in controversy for at least two hundred years. With new molecular testing of the morel (*Morchella*) genus, most agree there are perhaps eighty to one hundred species of morels worldwide, and many grow in the Western United States (Pilz 2007). In my mind, most Western morels fall into three broad groups based on color: whites, yellows, and blacks. Unfortunately, it's not that simple. Color is not always static. Morels can have gray, pink, green, or tan tones that show up at certain times in their growth cycle. Colors can also intergrade and change across habitats. Simply using white, yellow, and black colors to identify a morel species will not get you a positive identification. Still, most morels can be sorted into these basic groups, or *clades* (a term used to group organisms that have evolved from a common ancestor). Molecular evidence indicates that white, yellow, and black clades can contain one to several dozen species.

Gray morels in early stage of Rufobrunnea clade

Morels don't always fall simply into white, yellow, and black color schemes. Gray morels, like these, are commonly found, as well as pink, green, and tan tones for some species and developmental stages.

For simplicity's sake, let me describe the dominant white, yellow, and black clades in more detail and how they differ. But before I do that, you are probably wondering if you really need to know the difference between the clades. Heck no! But if you are as obsessed with morels as I am, you will find it fun and interesting to know the distinctions. And when you do, you can debate with your friends on such matters as to whether the mountain blonds taste better than the burn morels or whether there were more black morels this year than the last. But if you find these discussions boring, skip to the "Where Are They?" section.

The clades—*Rufobrunnea, Esculenta,* and *Elata*—also known as whites, yellows, and blacks, depending on your need to blabber scientific terms.

Rufobrunnea Clade (Whites)

The mountain blonds (morels) have vertically arranged pits, and the ridges of the pits are lighter than the pit.

In the Cascade Mountain Range, the mountain blond morels (also called blondies or whites) are a member of the *Rufobrunnea* clade. In this clade, the pits are generally arranged vertically up and down the cap, and the ridges of the pits are silvery white or buff, but never dark. The *Rufobrunnea* clade bruises brownish orange to pinkish where it has been touched—a characteristic for which the fungus is named. The shape of the cap is conical and often pointed at the top.

Another characteristic that distinguishes white clade from the *Elata*, or black clade, is the absence of a sinus—yes, morels can have a sinus—which appears as a "trench" at the connection of the cap to the stem. The mountain blond morel in the Cascades and the Western United States are mycorrhizal with true firs, pines, and perhaps other conifers. Globally, it is likely that some whites are also saprotrophic.

Esculenta Clade (Yellows)

The pits of the *Esculenta* clade, or yellow morels, do not line up vertically and generally occur in association with hardwood trees.

The *Esculenta* clade, or yellows, are yellowish-brown to light-brown morels that occur with hardwoods like madrones and cottonwoods. The ridges are buff and never dark at maturity. The pits are irregularly arranged, not vertically aligned like the mountain blonds, and the sinus is absent. The shape of the cap is rounded or somewhat "ovoid," not conical or pointy like the white clade. The yellow clade is primarily mycorrhizal with certain hardwoods in the Western United States.

Elata Clade (Blacks)

With the *Elata* clade, or black morels, the cap ridges are darker than the pits.

There is an indentation, or "sinus," where the black cap (right) connects to the stalk.

The *Elata* clade is a big group. Black morels, as their name implies, are dark brown (or even lighter when young) to black and have ridges that are darker than the linings of the pits. They have a conical or cylindrical top but not as pointy as the mountain blonds, and they are not rounded like the yellow clade. The pits are partially vertically arranged, and the sinus is always present. The black clade does not generally blush red.

In the Western United States, the *Elata* clade can be further divided into "natural blacks," "disturbance blacks," and "burn morels." The natural blacks form a mycorrhizal association with conifers that occur in undisturbed or natural settings. The disturbance black morels are associated with sites that have been impacted by soil compaction, logging, tree mortality, erosion, and other types of disturbances. They are often found along logging skid trails, downed wood, or sides of roads. The burn morel is associated with fire and are often found near areas that have been recently burned (one or two years following a fire), and it fruits in burn piles or areas impacted by wildfires. Burn morels are saprophytic and can be very abundant following fires in many areas of the Western United States.

A plethora of black morels from a burned area.

Where Are They?

Morels can occur everywhere and anywhere. Thirty-four morel species have been found in Europe, thirty-two species in Asia, and twenty-one species in North America (Pilz 2007). They are hunted internationally—Africa, China, Hawaii, Australia, Mexico, Canada, India, Israel, Java, New Zealand, Turkey, Sweden, Belgium, just to name a few countries. Wherever you travel, you will eventually run into someone who knows of a local morel source because morels are all over the fricking place. Yet ironically, they are often tough to find because they are camouflaged into the landscape and forest floor. Some species are global, while others are endemic to a specific habitat, region, or disturbance type. Morels occur from sand dunes to boreal forests. They are remarkable in terms of their "ecological elasticity," meaning they can be found in wilderness areas as well as cityscapes from the tropics to the boreal forests. For the saprophytic morels, a disturbance—such as fire, logging, tree death, soil disturbance, or the addition of landscape bark—can trigger abundant fruiting.

Morel spores travel far. The fire morel, for example, fruits prolifically over short periods of time (one to two years), and the spores they produce are light and can travel hundreds of miles. Humans also move morel spores around. This type of mushroom "trafficking," I believe, has been commonplace and widespread throughout human history. For example, the Italian Bronze Age iceman Otzi had three kinds of mushrooms in his possession when he was discovered after "chillin'" for thousands of years in glacial ice (Pientner et al. 1998). For humans traveling long distances, sun-dried morels would have been a logical food to pack, and they would have unsuspectingly spread morel spores along their journey. In addition, morel spores would have found fertile ground in areas where migrating peoples had created disturbances, such as burns and vegetative removal. It seems that morels, like humans (or perhaps because of humans), have achieved a certain level of global domination.

Humans and morels have developed a symbiotic and joyous relationship.

When Do They Occur?

Morels fruit following a cool winter period during springlike conditions. A general rule of thumb is that morels pop when snow has melted, soil is warming, and humidity is still high.

Abundant and prolonged fruiting requires persistent conditions of high humidity, rainfall, and warm temperatures. Often there are indicator plants that signal morels are fruiting. Different geographies have differing indicator plants. The flowering of *trillium* in the mountains of the Pacific Northwest is a well-known indicator. At lower elevations, it can be "when the lilacs are in bloom" or "when oak leaves are the size of mouse ears." The presence of certain cup fungi and false morels can be an indicator that morel fruiting is eminent. Morels emerge virtually full-sized from the ground, bent at a ninety-degree angle like a jackknife. They will grow a bit when they are above the soil surface if the conditions are favorable.

Morels generally emerge from the ground at a ninety-degree angle and are virtually full-sized.

In the Northern Hemisphere, early fruiting generally occurs at lower elevations and southern slopes, which are the first to warm up. As spring progresses, fruiting advances to the higher elevations and northern slopes, where fruiting can be as late as the beginning of summer. Hot temperatures and low humidity eventually terminate fruiting by drying out and shriveling the morels. At any location, the fruiting season typically lasts from two to four weeks before conditions are no longer favorable for morel production.

So good luck with your morel hunting. Hope "springs" eternal.

The size of the morel is not proportional to the size of the human that finds it.

Edibility

The sky is the limit for how you prepare morels for consumption. Morels are savory and delicious and can be a great substitute for meats in a recipe. But you should never eat the black, yellow, or white morels raw. They may contain trace amounts of compounds that cause gastric distress. With cooking, these compounds are vaporized. Morels can be dried whole or sliced longitudinally and dried. To use, rehydrate in water for thirty minutes until they regain their original size and concentrated flavor. If you don't have a drier, chop dry morels (do not wash), and stuff into baggies or a plastic container and place in the freezer. When you need them, simply unthaw and cook.

Similar Mushrooms That Should Be Avoided

As I will be discussing later in this section, there is a group of mushrooms that you may confuse with true morels because of their similarity in appearance. They are appropriately called "false morels" and include species such as *Gyromitra infula* and *Gyromitra esculenta*, elfin saddle (*Helvella lacunosa* and *H. vespertina*), and *Verpa bohemica*.

NATURAL BLACK MOREL

(Morchella elata clade)

The natural black morel mostly fruits on unburned soils or in undisturbed environments. The name originates from commercial harvesters who refer to the natural black being collected from conifer forests under "natural" conditions.

The natural black morel is also known as "black morel," "a natural," and *Morchella elata*.

Cap Characteristics

The cap is medium to large in size (generally three to six inches). When young, the coloration is steely gray to dark grayish brown turning black with age. The cap is honeycombed with pits and ridges. The ridges are generally darker than the pits. Vertical arrangement of the pits is highly variable. The cap is broadly rounded or ellipsoidal, not lobed or shaped like a brain or saddle, and it hangs over the stalk forming a sinus (a trench at the connection of the cap to the stem). It extends down, attaching nearly the length of the stalk when sliced longitudinally.

Stalk

The stalk is generally one to three inches long, is one hollow chamber, and ivory to light tan in color. It is smooth when young and turns somewhat grainy with age.

Fragrance and Edibility

Natural black morels smell earthy, woodsy, and nutty. They taste meaty and smoky and are highly prized by commercial and recreational pickers. (Probably a big reason why you bought this book.)

Habitat

Globally distributed, the natural black is associated with conifers in the Western United States. It mostly fruits in unburned soils or undisturbed "natural" environments. Like all morels, it blends into the forest and makes discovery challenging and rewarding. The natural black morels fruit when the winter snow is gone, the soil is warming, and the humidity is high. South slope and lower elevations fruit first in the spring. Higher elevations and north slopes fruit later in the spring or early summer.

The disturbance black morel is also in the *Morchella elata* clade and looks much like the natural black morel. The disturbance black morel occurs in areas where trees are in declining, dying, or recently dead. For example, massive disturbance morel harvests accompanied the Dutch elm disease 1971–1975 as it spread West across the United States. The disturbance black morel was reliably found at the bases of dying or recently dead elms. Compacted areas and areas where vegetation has been removed are also good habitat for the disturbance black morels. Declining or dying root systems seem to encourage fruiting of this fungus.

FIRE MOREL

(Morchella elata clade)

The fire morel often has a smaller and pointy cap.

The fire morel occurs on burned soils usually one or two years after the event. These can be areas of wildfire or slash-and-burn piles. The fire morel is usually smaller than the natural black morel. It is also known as "black fire morel," "burn morel," and "angusticeps."

Cap Characteristics

The cap is small to medium in size (one and a half to four inches). It is sometimes greenish when young and dark brown to black at maturity. The cap is honeycombed with pits and ridges. The ridges are generally darker than the pits, and the pits are not vertically arranged. It is an elongated, conic cap with a rounded apex, not lobed or shaped like a brain or like a saddle. The cap hangs over the stalk, forming a sinus, and extends down, attaching nearly the length of the stalk when sliced longitudinally.

Stalk

The stalk is generally one to three inches long and is one hollow chamber. It generally stays white with age, never brown. Smooth when young, it turns somewhat grainy with age.

Fragrance and Edibility

The fire morel has an earthy, nutty, woodsy fragrance. It is highly prized by commercial and recreational pickers and is meaty when cooked.

Habitat

The fire morel is common over the Western United States. It can be prolific in forest fire areas a year or two after the burn. It is found in burn piles, especially around the margins of the piles. Like all morels, it blends into the forest floor, charcoal, and burned debris, making discovery challenging and rewarding. The fire morels fruit after snowmelt, when the soil is warming and there is still high humidity. The massive fires in the Western United States have created short-term habitat for the fire morel. Commercial morel hunters study fire burn maps to chart their reconnaissance missions for spring-fruiting areas.

Fire morels in Alaska. How do you know you are having a good morel hunting day? When you must fill your pants with morels to get them all out of the woods.

YELLOW MOREL

(Morchella esculenta clade)

The yellow morel occurs in association with riparian trees and hardwoods generally at lower elevations. It is also known as "esculenta," "common morel," and "red-brown blushing morel."

Cap Characteristics

The cap is medium to large in size (generally three to seven inches). Occasionally, you can find a ten-inch cap. (You should be so lucky.) Its color is creamy white when young and turns rusty yellow or dingy, reddish brown with age. The cap is honeycombed with pits and ridges. The ridges are generally lighter than the pits, and the pits more rounded than elongated, not lined vertically up and down the cap. The shape is oval to subcylindrical or slightly tapered at the top. The cap is not lobed and not shaped like a brain or like a saddle. It is attached to the stalk and does not form a sinus. It extends down, attaching the length of the stalk when sliced longitudinally.

Stalk

The stalk is generally two to four inches long, hollow, and off-white to ivory or cream in color. The base enlarges with age with a pleated or gathered appearance.

Fragrance and Edibility

The yellow morel has a nutty, woodsy smell. Highly prized by commercial and recreational pickers, it is delicious fresh with all mushroom dishes. It is meaty and tasty in butter, in cream, baked, stuffed, and sautéed.

Habitat

The yellow morel mostly fruits in riparian environments and with hardwood trees such as ash, cottonwood, alder, madrone, and willow. It also occurs in oak forests and fruit orchards. It fruits at lower elevations generally when compared to the natural black and mountain blond morels. Prolonged fruiting requires persistent conditions that support morel growth, such as warming temperatures, rainfall, and humidity. It fruits early in the spring.

MOUNTAIN BLOND

(Morchella rufobrunnea clade)

The mountain blond occurs in association with mountain conifers on undisturbed or unburned soils. Other common names include "western blond morel," "blondie," and "white morel."

Cap Characteristics

The cap is medium to large in size (three to seven inches). In coloring, it is gray when young, turning to ivory with maturity. It bruises brownish orange and cinnamon to pinkish where it has been touched (a characteristic for which the fungus is named). The cap is honeycombed with pits and ridges. The pits are vertically arranged, and the ridges are generally lighter than the pits. The cap is columnar, pointy, and conical. It is not lobed or shaped like a brain or a saddle. It is attached to the stalk and does not form a sinus. The cap extends down, attaching the length of the stalk when sliced longitudinally.

Stalk

The stalk is generally two to four inches long, hollow, and ivory to tan in color. Smooth when young, it turns somewhat grainy with age.

Fragrance and Edibility

The mountain blond has a fresh fish, nutty, woodsy fragrance. It is **a** highly prized edible by both commercial and recreational pickers.

The mountain blond morel.

Habitat

The mountain blond is associated with conifers in the Western United States. It mostly fruits in unburned soils or undisturbed "natural" environments. Generally, it fruits just a bit later than the natural blacks in the Cascade mountains. Once you spot one, check your patch in earnest; they often occur in clusters.

THIMBLE MOREL

(Verpa bohemica)

The thimble morel resembles a true morel with its honeycombed cap. The key to identifying it from a morel is to slice it longitudinally and examine how the cap is attached to the stalk. If it is just attached to the very top of the stalk, it is *Verpa bohemica*, the thimble morel. The cap fits like a thimble on stalk's top. The sides of the cap hang off the top of the stalk like a skirt. But be careful bringing it to the dinner table. Some people have an adverse gastric reaction to eating the thimble morel if it is not cooked at a high temperature or if it is eaten in abundance.

The thimble morel is also known as "thimble cap," "wrinkled thimble morel," "early morel," and "early false morel."

Cap Characteristics

The cap is small to medium in size (one to three inches) and amber to yellow brown in coloration. The cap is wrinkled vertically and honeycombed like a true morel. The cap is not lobed and is shaped like a brain or a saddle. The cap fits like a thimble at the top of the stalk and hangs over it like a skirt. The cap does not extend down and attach to the stalk; it is only connected at the top of the stalk when cut in cross section.

Stalk

The stalk is generally two to five inches long, hollow, and white or pale yellow in color. It has light cottony stuffing in the stalk, and the flesh is brittle and fragile.

Fragrance and Edibility

The thimble morel does not taste as good as a true morel. Some people have an adverse reaction to it if it is not cooked at a high temperature or if they eat several. It may contain trace amounts of monomethyl hydrazine that cooks off at high temperature (rocket fuel).

Habitat

The thimble morel is found along streams, forest areas, or forest margins especially in the Pacific Northwest. It fruits in late winter and early spring, usually a couple of weeks before the true morels emerge.

FALSE MOREL

(Gyromitra esculenta)

The false morel is deadly poisonous when eaten raw. It is fairly easy to distinguish from a true morel, but if you are a beginner mushroom hunter, you will want to get to know the false morel and keep it away from the dinner table. The cap looks like a brain, and it is *not* honeycombed with pits like the true morel.

The false morel is also known as "brain mushroom," "lorchel," "elephant ears," "turban fungus," or "beefsteak morel."

Cap Characteristics

The cap is medium to large in size (three to six inches) and amber to reddish brown in coloration. The cap is very wrinkled but *not* honeycombed. It is very lobed and shaped like a brain.

Underside of Cap

The cap is connected to the stalk, so the underside of the cap is not visible.

Stalk

The stalk is one to three inches long, white, and primarily smooth, without prominent ridges. Usually, there are two or more cavities inside the stalk instead of one hollow cavity like the true morels. The flesh is brittle.

Fragrance and Edibility

Avoid consumption. Deadly poisonous when raw. The false morel has a fruity and nutty fragrance. *The Scandinavians and Eastern Europeans eat this mushroom with careful preparation and cooking.* In my view, it is not worth the risk even when it is parboiled or cooked to make it less dangerous. The active poison is called gyromitrin (for those few who passed college chemistry: N-methyl-N-formylhydrazine), which is metabolized to monomethyl hydrazine (or rocket fuel!) in the body. Gyromitrin is a toxin that destroys red blood cells in humans. It is toxic to the central nervous system and damages the liver and gastrointestinal tract. The name of this mushroom, *Gyromitra esculenta,* is very misleading. *Esculenta* means "delicious." As a guy who has helped name a few mushrooms, I find this unnerving. So if you find rocket fuel yummy and want to destroy your liver . . . this delicious *Gyromitra* is for you!

Habitat

The false morel is found in forest areas or forest margins at higher elevations. It usually fruits at a similar time as morels: spring and early summer. Collect true morels instead!

WESTERN ELFIN SADDLE

(Helvella vespertina)

The western elfin saddle is another odd-looking mushroom that has a cap shaped like a saddle and is not honeycombed like the true morel. Many reports indicate that elfin saddles can cause severe stomach upsets unless thoroughly cooked.

The western elfin saddle is also known as "elfin saddle," "slate gray elfin saddle," and *Helvella vespertina*.

Cap Characteristics

The cap is small to medium in size (one and a half to three inches) and has slate gray to black coloration on top. It is saddle-shaped, bald, and wrinkled. The edge of the cap is connected to the stalk in several sections. It is *not* honeycombed, and the flesh is thin and brittle.

Underside of Cap

The cap underside is gray to gray brown and bald.

Stalk

The stalk is two to six inches long. It is white to streaked gray and gray brown in color and is deeply ribbed and pocketed, extending up into the cap area.

Fragrance and Edibility

It is poisonous raw. The western elfin saddle's odor is not distinctive. It causes severe gastric distress with some individuals. Many reports indicate that elfin saddles cause stomach upsets unless thoroughly cooked. My advice—avoid elfin saddles since little is known about their short- and long-term effects; plus they have poor texture and taste.

Habitat

The western elfin saddle occurs under conifers in Northern California, Pacific Northwest, and the northern Rocky Mountains. Solitary or gregarious, it fruits in the fall or winter.

Recipe:

MOREL GNOCCHI WITH PEAS AND PROSCIUTTO

Chef James Daw

INGREDIENTS:

2 potatoes
1/3 cup flour
2 egg yolks
salt and pepper
4 tbs. dried morel powder
4 oz. of morels
1/4 cup of fresh peas
strips of prosciutto, julienned
1 shallot, diced fine
3/4 cup heavy cream
2 tbs. olive oil
1/4 cup white wine

METHOD:

Boil potatoes in salted water until tender, but not mushy. Drain and allow to rest until warm. Put through a ricer and add the egg yolks. Add flour and morel powder. Gently mix. Do not overwork the dough. Roll out on floured table, forming a log roughly 1/2 inch thick. Cut logs in 1/2-inch lengths to form gnocchi. Bring a pan of salted water to boil, and add the gnocchi for 1–2 minutes until they rise to the surface in several batches, allowing the water to return to a boil between cooking batches. Set aside when done.

Heat a 10-inch sauté pan over medium-high heat. Add olive oil and allow to heat up. Add morels and shallots. Cook for 3–4 minutes, then deglaze the pan with white wine. Cook for 2 minutes, then add the gnocchi. Cook for 2 minutes, and add the heavy cream. Reduce the cream for 3 minutes. Add peas and prosciutto, and season with salt and pepper. Garnish with shaved Parmesan or dots of goat cheese.

15. TRUFFLES

Humans have a long and rich history with truffles—a history as splendid and appetizing as truffles themselves. It started with the ancient Sumerians and Greeks, extended through the Renaissance, the French Revolution, and the World Wars. Today truffles are often produced in plantations and shipped worldwide to grace the tables of the finest restaurants on the planet. The truffle doesn't look like much: a small, potato-sized tuber that hides beneath the soil surface as a mycorrhizal associate on the roots of certain trees. But as inauspicious as these fungi may appear, they can fetch upward of a thousand dollars per pound, decorating the plates of the rich and famous and those looking for a unique culinary experience.

The French love the French black truffle even if this one came from a plantation in Australia.

The Western United States is blessed with hundreds of truffle species, which only a few native species are deemed to be of culinary value. The best, most abundant, and most expensive are the two Oregon white truffle species. The fall Oregon white truffle (*Tuber oregonense*) and the spring Oregon white truffle (*Tuber gibbosum*) are both commercially harvested. While the French black truffle (*Tuber melanosporum*) is native to Europe, it is presented in this section because truffle plantations are established in Pacific Northwest and because of their economic and culinary importance.

The two most notable truffles globally are the French black truffle and Italian white truffle from Southern Europe. The French black truffle is being produced in truffle plantations across the Western United States to varying degrees of success. *Fry, Thrive, or Die* includes the Oregon white truffles, collected from wildlands, that are commercially important and are a culinary pleasure when mature.

French black truffle (left) and Italian white truffle (right).

Historically, the French black truffles and Italian white truffles have come from wildland collections. However, an increased demand for the black truffle and decreasing wildland production have sparked an interest to grow them as a cultivated crop. The French black truffle is now being cultivated on the roots of hazels and oaks in truffle orchards of Northern California, Western Oregon, Washington, and Idaho. Over a hundred truffle orchards have been started in the Western United States, and while there have been varying degrees of success, several dozen orchards are now producing truffles. Australia, Italy, and Spain have established black truffle orchards with some yielding the French black truffle in great quantities (see the "Hazel Hill Truffle Orchard in Australia" essay). The Italian white truffle has never been cultivated successfully and relies exclusively on wildland collections in Italy and Croatia.

Truffles give off a pheromone-like molecule like a pig in heat, and for this reason, male pigs were historically used to sniff out truffles. The problem is . . . pigs enjoy truffles as much as humans, and they have been known to bite off the fingers of their handlers during truffle harvest. This is not a legend. In Italy, at a truffle hunting contest, I saw several truffle hunters with missing or partially missing fingers. This is one of the reasons why truffle hunters prefer dogs to pigs. But there is another reason too. Truffle hunters are secretive, so riding around with a dog in your car is an everyday scene. Have a pig in the back seat of your Fiat, on the other hand, and people are certain about what you are up to!

Pig digging for truffle.

Once a truffle treasure has been found, the hunter gently digs it up, being careful not to handle it too much. Fresh truffles need to be consumed within a couple days of harvesting since their flavor and scent are lost quickly. Rapid, careful handling and shipping are necessary to enjoy the full pleasures of the truffle.

Truffles grown in Australia for shipment to Hong Kong.

Truffles are best when used in dishes where their unique aromas can be absorbed in the starches or lipids of the foods being prepared. They are commonly grated over warm dishes like pasta, polenta, potatoes, and rice, but they are also magnificent with crab, lobster, scallops, and beef. The list of the foods that truffles can enhance is potentially endless. But be aware of how you store them. If you place truffles in the refrigerator next to items such as eggs, cream, butter, or rice, you may find that these items will also begin to smell like truffles.

A Very Brief, "Underground" History of the Truffle

- Aristotle, the ancient Greek philosopher, thought that truffles were a result of thunderbolts hitting the earth.

- Medieval Roman Catholic Church excluded the consumption of truffles because they believed eating truffles led to promiscuous behavior. The debate lingers on over whether there are truly aphrodisiac qualities to truffles.

- Napoleon and the French restored the truffle to its rightful place among the delicacies of exquisite meals. The French continue to celebrate its virtues.

- The golden age of truffles occurred in the nineteenth century. An insect (*Phylloxera*) devastated the vineyards of France, killing the vines. Truffle-bearing oak trees were planted in their place, resulting in huge truffle harvests. This was the golden age of truffles when truffle production flourished in the planted oak forests that dominated the landscape.

- The great decline happened when truffle production reduced dramatically in France and Southern Europe. Production in France has declined 96 percent from its peak in the late nineteenth century to the present (1883—1,500 tons; 1920—500 tons; and 2004—50 tons). Many factors led to the decline. These include habitat loss through two

World Wars; acid rains in Europe from 1950 to the mid-1970s; rural people stopped harvesting forest undergrowth for wood heat, creating more undergrowth, which is not conducive for truffle production; proliferation of non-truffle-bearing plants in traditional oak stands; and warmer, drier conditions associated with climate change. Many of these factors continue to threaten future wildland production.

- Truffle plantations started in the 1980s. Nurseries were established to produce the French-black-truffle-colonized tree seedlings. Roots of specific tree seedlings were inoculated with the truffle spores, and the resulting colonized roots of seedlings were planted into soils and landscapes specifically designed and managed to optimize truffle mycelial development. Truffle fruiting can take five to seven years, sometimes longer, sometimes not at all. Trained dogs assist in the harvest when truffles are ripe. Success has been mixed, depending upon the level of care, attention to soil, environmental conditions, and the expertise of the growers.

Oak trees in a truffle orchard.

OREGON WHITE TRUFFLES

(Tuber gibbosum and Tuber oregonense)

The Oregon white truffle grows below the soil surface in association with young Douglas fir forests on slopes below two thousand feet in elevation. The spring Oregon white truffle fruits from January to June, and the fall Oregon white truffle fruits from October to February. The soil and climate of the Northwest are similar to Piedmont and Tuscany in Italy and suitable for white truffle production. The Oregon white truffles get high marks for their culinary value. They are similar in aroma to the Italian white truffle but only when harvested mature. Indiscriminate raking of plots often yields an immature, nonaromatic truffles that have limited marketability and damages the reputation of the Oregon truffle market. Truffle hunting involves timing, technique, and experience. A skilled dog can determine when a truffle is ripe. That's when the odor of the Oregon white truffle is musky, spicy, fruity, savory, and smells like cheese. The same pheromone-like substance emitted by the white truffle to attract animals may have a similar effect on humans. Mature and aromatic Oregon white truffles sell to restaurants at $400 to $500 a pound—a quarter to half the price of the Italian white truffles, but still spendy!

The Oregon white truffle is also known as the "Oregon winter white truffle."

Exterior Surface Characteristics

The Oregon white truffles are small to medium in size (generally one half to three inches) and tan, buff, or brown in color. They have smooth surfaces without warts or ornamentation. The spring and fall Oregon white truffles look very similar and are largely differentiated by their different fruiting seasons.

Interior Characteristics

The solid interior is marbled with white veins. For the spring white truffle (*Tuber gibbosum*), the interior color is white when immature and becomes reddish brown to dark brown with maturity. For the autumn-fruiting white truffle (*Tuber oregonense*), the interior color is more reddish and orange brown with orange patches.

Stalk

It has no stalk.

Fragrance and Edibility

The smell of the Oregon white truffle intensifies with age and maturity. When it is mature, the fragrance is complex and difficult to define. Words such as *bright*, *spicy*, *garlic*, *onion*, *honey*, and *savory* have been used to describe the mysterious flavors and scents of the truffle. Fresh truffles need to be consumed within a few days of being harvested since their flavor and scent are quickly lost. All truffles are best when used in dishes where their unique aromas can be absorbed in the fats and starches of the foods being prepared. Popular herbs to pair with truffles are parsley, rosemary, thyme, basil, and oregano. Some chefs rank the Oregon white truffle, when mature, alongside the prized and expensive Italian white truffle.

Habitat

The white truffle is mycorrhizal and occurs in association with the roots of Douglas fir. Widespread in Oregon, especially in the coast range, it is most present in young to medium-aged Douglas fir that stands at lower elevations (below two thousand feet) and Christmas tree plantations. The spring Oregon white truffle fruits from January to June, and the fall Oregon white truffle fruits from October to February.

Similar Mushrooms That Should Be Avoided

There are hundreds of truffle species in the Western United States. Only a few are considered good edibles. None are known to be poisonous. A warning however: young *Amanita* species emerge from a volva, or egg sac, beneath the soil surface and can be mistaken by inexperienced mushroom collector as a truffle because of their shape and location.

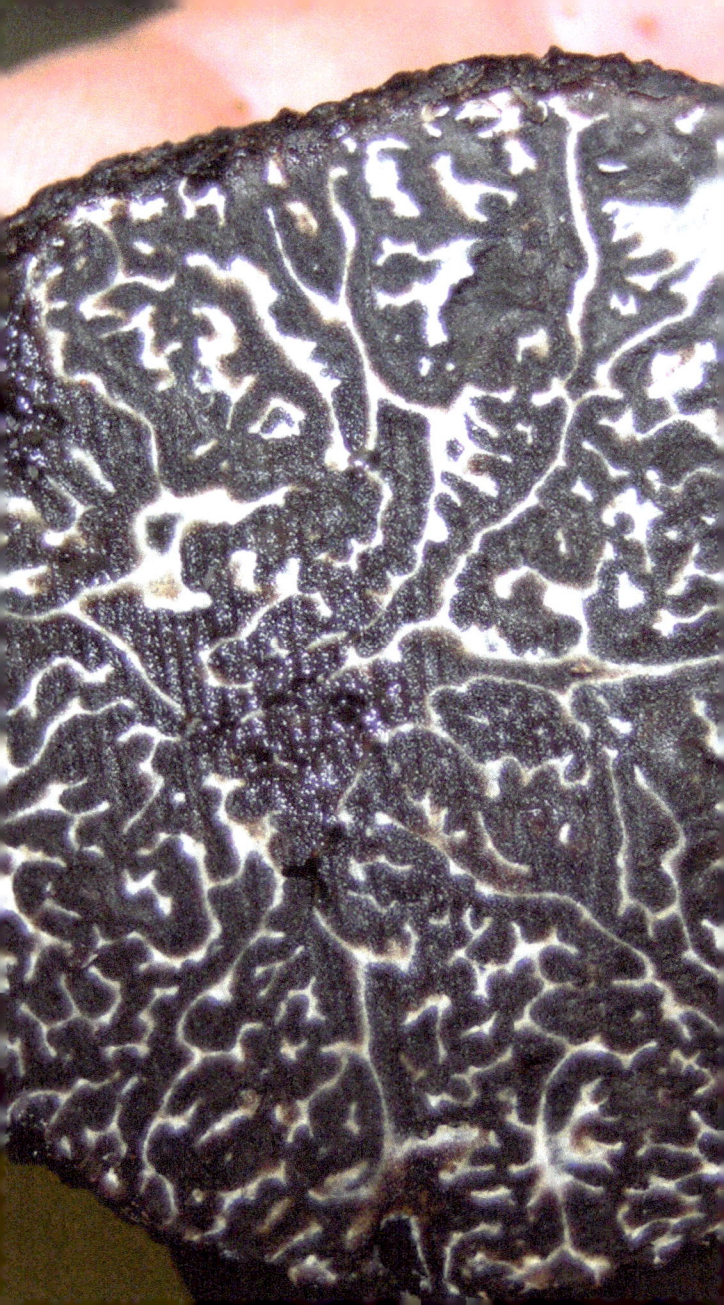

FRENCH BLACK TRUFFLE

(Tuber melanosporum)

French black truffle in cross section.

The French black truffle is native to Southern Europe and accounts for most of the native production in France, Spain, and Italy. Because of its high price and scarcity, it is being produced in dozens of plantations in California, Oregon, and Washington. Still in its early stages, truffle production in the Northwest is highly variable. The French black truffle has a dark-brown to black skin (peridium), which differentiates it from the Oregon white truffle, which is light brown. It also has warts or bumps on the outer surface, whereas the Oregon white truffles are smooth.

The French black truffle is also known as the "black diamond," "Périgord truffle," and "black truffle."

Exterior Surface Characteristics

Most French black truffles are two to five inches in size and generally round to oval in shape. The French black truffle is dark brown to black at maturity, with pyramidal "warts" or ornamentation on the surface.

Interior Characteristics

The interior is initially white, then darkens as it matures. Bright-white veins permeate the interior and turns darker with age.

Stalk

It has no stalk.

Fragrance and Edibility

The French black truffle has a strong odor. The smell of the truffle intensifies with age and is a complex, hard-to-describe combination of wet earth, pepper, dried fruit, and a hint of chocolate and sulfur. Some chefs rank the French black truffle as the epitome of culinary experience. It is used in dishes featuring fish, shellfish, pork, steak, cheese, pasta, and risotto. Eat freshly shaved over nearly cooked or warm dishes.

Habitat

The French black truffle is mycorrhizal and occurs in truffle plantations in the Western United States. It is associated with hazels, oaks, and other hardwoods.

Similar Mushrooms That Should Be Avoided

There are hundreds of truffle species in the Western United States. Only a few are considered good edibles. None are known to be poisonous.

HAZEL HILL: BLACK TRUFFLE PLANTATION EXTRAORDINAIRE

Dr. Nick Malajczuk with the French black truffle in Australia.

In the 1990s, Australian mycologist Dr. Nick Malajczuk had an ingenious idea: grow French black truffles in the Southern Hemisphere. His vision was to produce fresh truffles from June through August in the Australian winter, exactly when they were in short supply in the Northern Hemisphere. The reason he thought this would work is because Australia is dominated by eucalyptus forests, which have not evolved to host-competing fungi that could limit French black truffle production. Nick searched for areas with climates like parts of France where black truffles grew in abundance and eventually settled on land near a small town in Southwest Australia called Manjimup. It was nestled among orchards, wineries, and picturesque rolling hills that contained stately karri and jarrah eucalyptus forests.

Nick found a few investors to support his venture and purchased fifty acres that he called Hazel Hill. He began by growing a range of oak and hazel seedlings in a local greenhouse, inoculating them with a variety of black truffle spores he had gathered from France. Then in 1997, Nick began planting the orchard on a grid and carefully installed experiments with different treatments. He wanted to understand if he would see a difference in the level of truffle production if he changed the structure, chemistry, and other characteristics of the soils. He tried a variety of experimental treatments that included soil amendments, soil moisture regimes, cultural practices, and canopy management. Fortunately for me, I was flown over from the States on several occasions to help monitor the root systems of the planted trees for the presence of the black truffle mycorrhizae. We found that some treatments really encouraged mycorrhizal proliferation, while others did not. In 2003, the first truffle was discovered by Nick's trusty dog named Guinness. It was a great celebration years in the making. A triumphant feast followed.

Hunting black truffles with a dog at Hazel Hill, Australia.

Over time, the most successful treatments were implemented across all of Hazel Hill. In the summer of 2006, Hazel Hill produced a whopping 1,300 pounds of black truffles, which found their way fresh to restaurants around the world at a season never available previously. Working at Hazel Hill was a mushroom experience of a lifetime. We would walk up and down the rows of trees with a dog, flagging the areas the dog sniffed and scratched, which would indicate that a truffle lay beneath the surface. Later we would return to the flagged areas with a small trowel, smell the earth for the black truffle scent, and pop the black truffle from the ground. After our bags were full, it was into the processing barn for weighing, cleaning, and shipping.

Black truffles as black bumps in the soil at Hazel Hill.

In 2014, black truffle production at Hazel Hill exceeded twelve thousand pounds. Hazel Hill had become perhaps the most productive truffle plantation in the world. Now called Truffle Hill, the site sells a range of products beyond fresh black truffle, such as truffle oil, truffle aioli, truffle tapenade, truffle honey, truffle salt, and truffle risotto. It also allows a limited number of people each year the chance to enjoy a truffle hunting experience and learn how to find mature truffles that are ready for harvesting. With the success of Truffle Hill, dozens of truffle plantations have sprung up in the southernmost regions and on the east and west sides of the Australian continent. Most of these plantations are producing black truffles, but none have produced close to the quantities seen at Truffle Hill.

ITALIAN TRUFFLES ARE PRETTY, PRETTY GOOD

Jack Ingvaldson

We had traveled to Alba, Italy, for the famous annual fall harvest festival celebrating the Italian white truffle *Tuber magnatum*. My wife, Kim, and I were in the company of Dr. Mike Amaranthus; his wife, Eileen; and Dr. Jim Trappe, the recognized godfather of mycology at Oregon State University and a world expert on truffles. (Jim has discovered over a hundred new truffle species.)

Alba, Italy, is the sister city to Medford, Oregon, our neighboring city; so when we arrived, we were met by the Alba mayor and treated like royalty. We were hosted to a lunch, which gave us a hint of what was to come. Featured was beef with shaved white truffles. Accompanying that was risotto infused with truffle oil and topped with more shaved truffles. Of course, a nice chianti went well with lunch. Then there was truffle dessert and dessert wine. Our conversations with the mayor and others centered around—what else?—truffles. I was thinking this mycology gig was pretty good.

The following day, we toured the city and took in all that the festival had to offer. A parade proceeded through old town, highlighted by a group dressed up as their favorite mushrooms. There were several porcini guys, several marching lumpy truffle tubers, and a lone Caesar's amanita. Shortly thereafter, we arrived at a large circus tent where perhaps forty truffle vendors displayed their recent harvest. The first thing I noticed was the incredibly strong odor of the fresh Italian white truffle. Having never experienced such a powerful aroma, I was taken aback. It was like a spicy, earthy bomb had just gone off when I entered the tent. Every table featured the white truffle next to medical grade scales, which were there to accurately weigh the most expensive condiment in the world at $295 an ounce. Chefs from all of

Europe were in the tent buying truffles to feature at their restaurants.

The following day, we were greeted by two Italian university professors who had studied under Jim at Oregon State University. They had arranged for us to have dinner at the Il Cortile Trattoria in Parma, Italy. As we entered the restaurant, we were seated by the maître d' holding a basket full of fresh black truffles for the meal. The truffles had come from a truffle orchard that one of the professors had helped establish near Parma. He had arranged for them to be sent to the restaurant to celebrate Jim's visit to Italy.

Grating the black truffle over a hot entrée.

What followed was the most extraordinary three-hour meal (showcasing freshly shaved truffles at quantities we had never experienced) and a six-course dinner featuring truffles shaved on each course. Numerous bottles of Italian Barolo accompanied each course. We left totally in awe of the most incredible gastronomic experience of a lifetime. Again, I was left thinking this mycology gig is . . . pretty, pretty good.

And appreciating anew what I had often heard Dr. Jim Trappe proclaim, "Why would anyone pick a profession other than mycology?"

TRUFFLES, TREES, AND CRITTERS

A Douglas squirrel gobbles a *Rhizopogon* truffle

Truffles are the fruit of a network of underground filaments called *mycorrhizae*. These filaments act as an extension of the tree root system, attaining water and nutrients in exchange for sugars from the plant. The fruit of the underground mycelial network is the truffle, and it is packed with spores, the "seeds" of the truffle.

Truffles develop entirely underground, unlike their close aboveground relative—the mushroom. Mushrooms typically disperse their spores in the wind—in sharp contrast to truffles that need the help of animals to move their spores around. Most truffles, including the Oregon white truffles, are eaten by mammals that dig and gobble them up. When you walk in the woods, you have probably seen thousands of small animal "digs," which are the shallow pits in the forest duff where truffles have been excavated by wildlife. The vast majority of these truffles are not species humans value for culinary value but are critically important to the diets of wildlife species in the Western United States. Human truffle hunters use these animal digs as clues to where the Oregon white truffles may be hiding. Truffles have evolved to emit powerful aromas that appeal to animals and signal that they are mature and ready to be dispersed. Truffle spores pass through the animal digestive system without impact and are ready to form a new mycorrhizal colony when deposited at some distance from the original meal. Truffles are an important food source for many wildlife species not only in the Western United States but also in forest ecosystems all over the world.

16. CORALS AND CONKS

In 1928, at Saint Mary's Hospital in London, Western medicine found its most valuable botanical of the century. It was a fungus: penicillin. This discovery has saved millions of lives. Today replicating the remarkable discovery and use of penicillin is highly unlikely, but interest in fungi in clinical research is rapidly increasing. The six corals and conks in this section have potential medicinal value. Medicinal mushrooms have a long history of traditional use, and recent science has demonstrated significant therapeutic benefit for a range of illnesses and conditions. Mushrooms are an exciting area of natural health, and for forty-plus years, I have seen improved clinical research studies and peer-reviewed evaluations. However, to fulfill its potential, many questions remain to be answered. As they say, Rome was not built in a day!

I discussed corals and conks together because of their potential medicinal properties. Coral fungi are multibranched, coral-looking fungi that arise from a fleshy base. Two coral fungi species—the hen of the woods and the cauliflower mushroom—are discussed in *Fry, Thrive, or Die*. Both are both medicinal and delicious edibles.

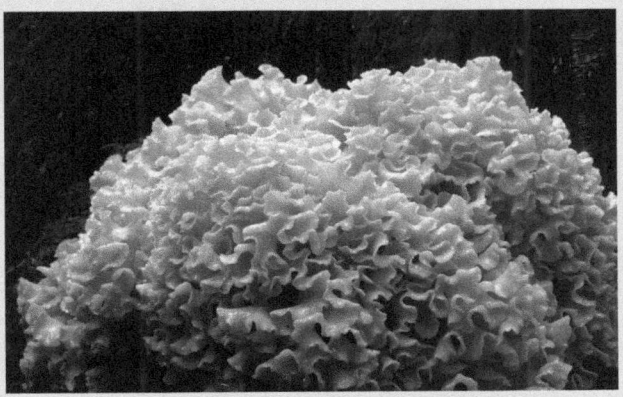

The cauliflower mushroom.

Conks are those tough, woody bracket, or shelflike fungi you see growing on dead trees and stumps. They are an important part of the ecosystem, removing the dead organic debris that would pile up in forests if they were not present. They are not soft enough to eat, but many have been used medicinally as teas and powder in Asia for millennia. Recent research on the active ingredients in many conks have been driving the use of these fungi to treat a range of illnesses. Although more human trials are necessary, the unique, biologically active compounds in many conks are now being appreciated and utilized by Western medicine. The five conks covered in this section are turkey tail, varnish conk reishi, artist conk reishi, red belt conk, and chaga.

The artist conk reishi.

HEN OF THE WOODS

(Grifola frondosa)

The hen of the woods is a coral fungus known for its culinary and medicinal value. They are the beautiful and distinctive multibranched, polypore that fruits at the base of large hardwoods. There are no look-alikes that are toxic.

The hen of the woods is also known as "maitake," "ram's head," and "sheep's head."

Form and Size

It grows from an underground tuber-like structure into a tight, multibranched cluster. The hen of the woods consists of multiple branched caps that are curled in the shape of a spoon with wavy margins. Underneath the multiple caps are minute pores and a thick, dense central stalk. Generally moderate to large size, it occasionally can be massive.

Color

It has grayish-brown, multibranched caps and milky-white central stalk.

Fragrance and Edibility

The hen of the woods has an earthy and peppery smell. It is savory and delicious and one of the best edible mushrooms. This culinary delight has been enjoyed by the Japanese for millennia. Add clean and sliced hen of the woods to meat, fish, pasta, stir-fry, and soup dishes. It can be marinated, sautéed, barbecued,

stir-fried, or baked. It is a popular dried medicinal mushroom powder and taken in capsule form.

Habitat

Hen of the woods is common in Asia and the Eastern United States but is also found in Oregon, Washington, and Idaho in the late summer and fall. Hen of the woods is also being cultivated in the Western United States, where it is grown on alder sawdust and oat bran. It is sometimes found in growers' markets and gourmet groceries. It fruits in the same hardwood log or stump year after year.

Medicinal Value

Hen of the woods is getting considerable attention from the pharmaceutical industry for its medicinal benefits. It has long been used as a medicinal mushroom in Japan and Korea. It has a high level of antioxidant activity (Postemsky and Curvetto 2013). Maitake also has a strong antitumor effect (Masuda et al. 2013). Trials are underway for treatment of breast, lung, and liver cancer. The causal compound is most notably beta-glucan, which can compose 10 to 50 percent of its dry weight (Masuda et al. 2015). The hen of the woods (*Grifola frondosa*) is popular as a powder mixed in capsules and medicinal products.

THE CAULIFLOWER MUSHROOM

(Sparassis radicata)

The cauliflower mushroom is a coral fungus known for its culinary and medicinal value. It is a distinctive multi-branched polypore that fruits at the base of conifers in the Western United States. The mass of flattened lobes gives this mushroom an appearance of tightly packed egg noodles or cauliflower. There are no look-alikes that are toxic.

The cauliflower mushroom is also known as *Sparassis crispa*.

Form and Size

The cauliflower mushroom's multibranched, flattened lobes are relatively tough. The central stalk is thick, dense, and often extends deep into its rooting base. It ranges from large (one to five pounds) to occasionally very large (greater than forty pounds!).

Color

The color is white to yellowish when fresh and turns tan with age.

Fragrance and Edibility

The cauliflower mushroom has an earthy and almondy smell with a hint of morel fragrance. Delicious but chewy, it can be made more tender by slow cooking. It is a popular edible and can be sautéed, baked, or added to soup. Sometimes it is difficult to clean. It stores well fresh and is resistant to decay. It can also be dehydrated and reconstituted.

Habitat

The cauliflower mushroom is widespread in conifer stands across the Western United States and common in California and the Pacific Northwest in particular. It fruits in the same conifer log or stump year after year.

Medicinal Value

The cauliflower mushroom is widely used in traditional Chinese medicine and contains a high concentration of certain beta-glucans of known medicinal value. The dry weight of the cauliflower mushroom was found to contain 43 percent beta-glucan, which is approved for cancer treatment in Japan (Sharma 2022). Experiments suggest the cauliflower mushroom contains chemicals that stimulates the immune system and has biological properties that are antitumor, anti-inflammatory, and antiviral (Ohno et al. 2000, Chandrasekaran et al. 2011).

Similar Mushrooms That Should Be Avoided

A larger group of coral mushrooms species called *Ramarias* are not included in *Fry, Thrive, or Die*. They are described in the David Arora's *Mushroom Demystified*. The *Ramarias* are a large complex group that are still being sorted out taxonomically. Some *Ramarias* are edible, while some are slightly poisonous, and distinguishing between the species can be difficult. The medicinal value of the *Ramaria* group is also unknown. However, they are always a welcome sight on a foray—beautiful and mysterious entities pushing up through the forest floor.

TURKEY TAIL

(Trametes versicolor)

Although too tough to eat, turkey tail is popular as a medicinal fungus when prepared in teas, tinctures, or capsules. True to its name, the fruiting body resembles a turkey's tail. It is distinctive with no toxic look-alikes.

The turkey tail is also known as *Coriolus versicolor* and *Polyporus versicolor*.

Form and Size

The turkey tail ranges from small to medium in size. It varies from shelflike shapes, multiple rosette shapes, to fan shapes. Zoned conspicuously with varying colors, it often has velveting, hairy zones. A stalk is absent or minimal. The pores are minute, thin, and vary from white to very pale brown. It does not stain when handled. Turkey tail is tough like leather when young and hard and rigid when old or dry.

Color

The color is highly variable from gray and charcoal to cinnamon, red, and brown in alternating zones.

Fragrance and Edibility

Fresh young turkey tail has a mild fruity smell. It is bitter to taste and too tough and leathery to eat. It is a popular medicinal dried mushroom powder and taken in capsule form and available as a tincture.

Habitat

Turkey tail is found all over the world. It is especially common on oaks and other hardwoods in the Western United States. It rarely occurs on conifers. This saprobe breaks down and lives off dead wood. In this important ecological role, it helps clean the forests of organic matter and create space for new growth. It is common and often present in clusters on hardwood stumps, down logs, and branches and is harvested year-round.

A single log or stump can have over a hundred turkey tails.

Medicinal Value

Turkey tail is one of the best-documented medicinal mushrooms and is a source of chemical compounds with anticancer activity (Tatiana et al. 2020). It contains PSK (polysaccharide Kureha) also known as krestin, which is approved for use as an anticancer agent in Japan. PSK extract from turkey tail is a frequently used cancer treatment drug in Asia and is being tested on humans in Western countries (Nakazato et al. 1994). The FDA recently approved a clinical trial for turkey tail extracts, allowing patients with advanced prostate cancer to take it in combination with conventional chemotherapy. Numerous human studies have been published on turkey tail extract for gastric, lung, and breast cancer. To access fourteen recent scientific studies on turkey tail medicinal properties, go to mushroomreference.com. A wide variety of turkey tail products are available commercially as teas, tinctures, capsules, or as an ingredient in coffee and other drinks. Fruiting bodies are generally ground to a fine powder or broken into one- to two-inch chunks used to make tinctures.

VARNISHED CONK REISHI

(Ganoderma oregonense)

The varnished conk reishi grows as a bracket fungus on logs, stumps, and dead standing conifers. As with many *Ganoderma* species, the varnished conk reishi is valued for its medicinal properties. There are no toxic look-alikes, but the varnish conk reishi should be harvested before it decays or gets moldy.

The varnished conk reishi is also known as "western varnish conk," "Oregon reishi," and "Oregon ling chi."

Form and Size

The varnished conk reishi is large and shaped like a hoof or a bracket, with no stalk or stubby lateral stalk. It has a hard outer crust, and it is soft beneath the crust.

Color

It has a smooth, varnished dark reddish brown to mahogany cap. The pore surface underneath the cap is white to yellowish white and stains brown when scratched.

Fragrance and Edibility

The varnished conk reishi has an earthy smell and is bitter and too tough to eat. It is popular dried medicinal mushroom powder and taken in capsule form and as a tincture. It is also a medicinal tea when dried chunks are boiled in water for five minutes. Fruit juice or other beverages can be added to improve the taste.

Habitat

Parasitic and saprophytic, the varnished conk reishi grows on dead standing conifer trees, logs, and stumps. It occurs in conifer stands across mountainous areas in the Western United States. It fruits in the fall to midwinter.

Medicinal Value

The varnished conk reishi is a relative to *Ganoderma lucidum*, which has been used for millennium in Chinese medicine. The Chinese believe that reishi promotes good health, longevity, aids digestion, and prevents cancer. Published studies are indicating a strong antitumor effect (Tank et al. 2006, Tank 2006). Reishi is widely available from health food stores in tinctures and capsules. It is also an ingredient in many medicinal mushroom products.

ARTIST'S CONK REISHI

(Ganoderma applanatum)

The artist's conk reishi has been used as a natural artist's canvas for years. Under the cap is a white-pored surface that turns permanently brown when scratched. Artists have made etchings on these surfaces, often of nature scenes, and sold them in novelty shops and art studios. Unlike the varnish conk reishi, this conk does not have a shiny or varnishy cap.

The artist's conk reishi is also known as "artist's bracket," "shelf fungus," and "tree tongue."

Form and Size

The artist's conk reishi ranges from large to very large. It is shelf-like, has no stalk, and is hard and solid. The surface is furrowed and ridged.

Color

The cap is brown to gray, not shiny. The pore surface underneath the cap is white and stains brown when scratched. This can be used by artists as a "canvas" for their drawings. The tiny white pores under the conk turn brown on mature or overmature specimens.

Fragrance and Edibility

The artist's conk reishi has an earthy smell. Obviously, it is too woody to eat. It is a popular dried medicinal mushroom powder and taken in capsule form or as a tincture. It is also a medicinal tea when dried chunks are boiled in water for five minutes.

Habitat

Parasitic and saprophytic, the artist's conk reishi grows on stumps, downed wood, and dead standing hardwoods. Occasionally, it is found on mature and old-growth Douglas fir. Widespread and common across the Western United States, its perennial fruiting body expands in the fall and early winter. Artist's conk can live for decades.

Medicinal Value

Reishi is widely available in health food stores as a supplement in powder, capsules, and tinctures. It is a popular ingredient in other products like coffee, teas, and chocolate. Field specimens of the artist's conk reishi can be turned into one- to two-inch chunks with a hatchet or saw and used in the preparation of teas or tinctures. The tea is slightly bitter, and fruit juice can be added to improve the taste. Active ingredients include antioxidants, antivirals, antitumors, triterpenoids, ganoderic acids, and beta glucans (Mohammadifar et al. 2020).

Reishi: A Rich Medicinal History and Use

Asian reishi (*Ganoderma lucidum*) is a close relative and has been used medicinally for two thousand years by the Chinese (Jones 1990). Writings from the Eastern Han Dynasty in China (AD 25–220) describe the beneficial effects. Images of reishi in Chinese art, furniture, and carvings began in AD 1400. Records from the Ming Dynasty AD 1590 indicate reishi was used to enhance vital energy and strengthen cardiac function and memory. Reishi in China was also believed to have antiaging effects, which gave it the name "mushroom of immortality." Its benefits, combined with its scarcity, made the use of reishi expensive. Only the rich and privileged of Chinese society could afford the fungus. Today reishi is a popular medicine throughout Asia, and its use is growing in North America. Recent scientific studies (fifty available at mushroomreferences.com) document the beneficial effects of reishi for cardiac health, liver health, and its antioxidant, antitumor, and antimicrobial activities.

CHAGA

(Inonotus obliquus)

Chaga is too tough to eat and ugly in appearance but popular as a medicinal fungus. You might walk right by it because it resembles a clump of burnt charcoal on select hardwoods.

Chaga is also known as "clinker," "birch canker polypore," and "cinder conk."

Form and Size

Chaga ranges from medium to large in size. The sterile conk grows on birch trees and a few other hardwoods like poplar and alder. It resembles a clump of burnt charcoal, but inside the tough outer surface is a soft orange core.

Color

The color is charcoal brown to black on the exterior and brown and reddish brown interior.

Fragrance and Edibility

Chaga has a mild sweet, earthy, aromatic smell. The smell has hints of vanilla. It is too woody to eat.

It is a popular dried medicinal mushroom powder and taken in capsule form and as a tincture. It is also a medicinal tea when dried chunks are boiled in water for five minutes. It has a mild and pleasant flavor.

Habitat

Chaga occurs naturally in higher northern latitudes primarily on birch trees. It is also found on birch ornamental plantings in urban and suburban areas.

Medicinal Value

Chaga is among the greatest sources of natural antioxidants and powerful immune support (Balandaykin and Zmitrovich 2015). The mushroomreferences.com website has twenty-one recently published studies on the medicinal properties of chaga. Its high level of anti-inflammatory properties makes it a potential alternative remedy for maladies such as arthritis, high blood pressure, and high blood sugar levels. Chaga has very high concentrations of melanin, which is present in our hair, eyes, and skin. Melanin from fungal sources can improve the health of skin epidermis and maintain skin and hair pigmentation. It is also what turns chaga tea water brown when brewed. Beneficial compounds in chaga may also help slow the growth of cancer cells (Kim et al. 2006, Sun et al. 2014). Several studies are underway investigating the ability of chaga extracts to treat breast, lung, and colorectal cancer. While some studies have produced promising results, more human trials are needed. A wide variety of chaga products are available commercially as teas, tinctures, capsules, and as an ingredient in coffee and other drinks. Fruiting bodies are generally ground to a fine powder or broken into one- or two-inch chunks (chaga chunks) used to make teas and tinctures.

How to Make Tinctures

It is easy to make tinctures from dried turkey tail, chaga, and reishi mushrooms. Tinctures are an extraction of both alcohol- and water-soluble compounds. Break your conks into one- to two-inch chunks, or grind them to a fine powder. Pack a glass jar (don't use plastic) with your mushrooms, and cover with high-percentage alcohol, like Everclear or vodka. Seal the container, and shake or stir once a day for two to four weeks. When you are ready, decant the infused alcohol through a filter cloth, and squeeze the liquid out of the remaining mushrooms through the filter cloth to extract all the alcohol. For the water extract, place your one-part mushrooms in a saucepan with twenty-parts water and boil for five minutes. Allow the water extract to cool, and add it to the alcohol infusion at a 5:1 ratio. Tinctures are stable and can be kept for at least a year at room temperature. You can add two full droppers of tincture to water, juice, coffee, or any other beverage morning and night.

RED-BELTED CONK

(Fomitopsis pinicola)

The red-belted conk is shelflike or hoof-shaped; it is very hard. Medium to large in size, it's colored red, reddish brown, or cinnamon bands and brown or black where it connects to the tree. It has a pale-yellow band at the margin of the conk. It does not have a varnished cap and has a white pore surface underneath the cap that does not stain when scratched. There is no stalk.

The red-belted conk is also known as "red-belted fungus," "red belt bracket fungus," "red-banded polypore," and *Fomes pinicola*.

Form and Size

The red-belted conk ranges from medium to large. It is shelf- or hooflike, with no stalk, and is very hard and solid. The surface of the conk is furrowed and ridged, with each "ring" representing a year's growth.

Color

The color is reddish to cinnamon and brown to black where it connects to the tree. The cap is dull, not shiny, and the outer edge of the conk is often a pale-yellow color. The pore surface underneath the cap is white and does not stain brown when scratched.

Fragrance and Edibility

The red-belted conk is obviously too woody to eat. Less research has been done on the medicinal value of the red-belted conk, but it does contain many of the same active ingredients as reishi. Field specimens of the red-belted conk are very hard, and with elbow grease, it can be turned into one- to two-inch chunks with a hatchet or saw and used in the preparation of teas or tinctures. The tea is slightly bitter; fruit juice can be added to improve the taste. Active medicinal ingredients are antidiabetic and anti-inflammatory with antioxidant-supporting compounds (Zahid et al. 2020, Onar et al. 2016).

Habitat

Growing on stumps, downed wood, dead standing conifers, and occasionally hardwoods, the red-belted conk is widespread. It is very common across the Western United States from the Rocky Mountain Range, Cascades Mountain Range, Sierra Nevada Mountain Range, and coastal ranges of California, Oregon, and Washington. Its perennial fruiting body expands in the fall and early winter. The red-belted fungus can live for decades.

SPECIAL DEDICATION TO DR. JAMES M. TRAPPE, MYCONAUT

Dr. James Trappe

Sixty years ago, astronauts commanded missions to unlock the secrets of space. Scientists worked together to travel to the moon and discover the wonders of the universe. Recently, we celebrated the fiftieth anniversary of humans' first walk on the moon.

Sixty years ago, the fungal research world also set off on a journey. A journey to unlock the mysteries of a living universe right beneath our feet. A journey of discovery all of us share in *Fry, Thrive, and Die*.

Dr. Jim Trappe was one of our first fungal astronauts—a "*myconaut*"—under whom I had the honor of studying at Oregon State University and the United States Forest Service. Jim introduced me to the mysterious universe below the surface of the soil and how it influences the ecosystems we live in. I had the opportunity to travel around the world with Jim in search of mycorrhizae, truffles, and mushrooms . . . celebrating the cultures and gastronomic pleasures that followed fungi. Jim had friends on every continent.

Dr. Jim Trappe (left) collecting fungi with Dr. Gaston Guzman (center), a pioneer of *Psilocybe* research, in 1976 in Veracruz area of Mexico.

Jim has published scientific papers across seven decades. Now over ninety years of age, Jim is still writing scientific manuscripts. His résumé is unreal. After receiving his PhD in forestry in 1962 from the University of Washington, Jim began a long and distinguished career with the United States Forest Service Research Station as project leader for the Mycorrhizal Research Team in Corvallis, Oregon. With over five hundred mycorrhizal scientific publications dating back to the 1950s, his work in fungal taxonomy and mycophagy has been groundbreaking. Over the course of his career, Jim has been author or joint author of 1 new order, 3 new families, 42 new genera/subgenera, 215 new species, and 168 new species combinations. Jim has 18 fungi named in his honor.

Dr. Jim Trappe with his trusty "truffle fork."

Few, if any, scientists living today have earned Jim's taxonomic prowess. Besides classifying fungi, Jim has been interested in the role fungi play in forest ecosystems. With other scientists, he discovered that fallen trees (those pesky obstacles that get in our way while mushroom hunting) are critical for the long-term sustainability of a forest. At another point in his career, he and a wildlife biologist studied the relationship between truffles and the flying squirrel and a number of other forest animals that depend upon fungi for food and survival. They showed that animals and fungi can be intricately interconnected, a finding that enlightened many of us to the importance of maintaining old-growth forests.

Decades ago, Jim—along with a few other scientists, including me—made several trips to Mexico, China, Europe, Australia, and Thailand to locate and describe new fungal species. One particular foray took us to Tasmania, Australia, to collect and classify belowground-fruiting fungi. On this trip, we visited a rural forest area called Cradle Mountain where we hit a fungal "mother lode." In this remote landscape, we discovered dozens of new truffle species and even several new genera and families of fungi.

Dr. Jim Trappe taking notes on truffle specimens in Tasmania.

While this was pretty amazing in itself, what was really cool was our observation that so many Australian marsupials were also foraging for these same truffles. Australian possums, kangaroos, wallabies, bettongs, potoroos, bandicoots, and wombats were in search for and depended upon these belowground-fruiting fungi as a food source.

An Australian brushtail possum finds a fungal treat.

A common wombat wanders across Tasmania.

After evening meals at the local restaurant, we would watch the cooks throw food scraps off the back deck to feed the wildlife. We would add the excess truffles from our daily collections to the bounty. Mountain brushtail possums, wallabies, potoroos, wombat, and bandicoots would jockey for position to eat the goodies, to Jim's great delight.

One night there was a conspicuous clatter of strange noises after the feeding. A Tasmanian devil had slowly wandered into the middle of the smorgasbord. All the other animals had backed up into the shadows while the devil sat perched on top of the food, slowly spinning to protect his food stash, like the Taz character in a Warner Brother's cartoon. You should have seen the twinkle in Jim's eye.

For seven decades, Jim has generated unique passion, insight, and humor to the fungal community. He is respected and loved by the hundreds of students, visiting scientists, and friends he has influenced and nurtured over his remarkable career. Jim is a myconaut for the ages.

Jim shares the fragrance of the Italian white truffle in Italy.

THE AMAZING FUNGAL AND ANIMAL CONNECTION

Dr. Jim Trappe

I've been amazed over my career how clever and intertwined animals and fungi are. Fungi use animals to disperse their spores in a host of different ways. One of my favorites is the flying squirrel. It descends from the forest canopy at night, attracted by pheromones produced by the underground-fruiting truffles. It consumes the truffle, climbs back into the forest canopy, and sails from tree to tree across the forest landscape.

A flying squirrel descends from the forest canopy in search of truffles.

A flying squirrel eats a truffle.

Inside the little belly of the flying squirrel, digestive juices break down the beneficial proteins, vitamins, and minerals of the truffle. But the truffle spores remain viable. The ornamentation of the spores and their hydrophobic nature protect the spore wall from digestion. As the squirrel flies from tree to tree, it poops the spores down to forest floor, where they can germinate and form new mycorrhiza and eventually new truffles.

This poop rain has been going on for millions of years. Any book written by Dr. Mike Amaranthus needs a healthy dose of discussion of poop and gastric juices. Now you have it.

REFERENCES AND WEBSITES

These are handy references and websites for more information about frying, thriving, and *not* dying while using mushrooms.

Field Guides to Mushrooms
- *Mushrooms Demystified* by David Arora
- *Mushrooms of Cascadia* by Michael Beug
- *National Audubon Society Field Guide to North American Mushrooms* by Gary Lincoff

Cookbooks
- *Fantastic Fungi Community Cookbook* by Eugenia Bone
- *Edible Mushrooms: Safe to Pick, Good to Eat* by Barbro Forsberg
- *Cooking with Wild Mushrooms* by Ingrid and Pelle Holmberg
- *The Mushroom Cookbook* by Michael Hyrams and Liz O'Keefe

Psychedelic Mushrooms
- *Psilocybin Mushrooms of the World* by Paul Stamets
- *How to Change Your Mind* by Michael Pollan

Medicinal Mushrooms
- *Medicinal Mushrooms of Western North America* by Robert Rogers and J. Sept
- *Medicinal Mushrooms: A Clinical Guide* by Martin Powell
- *Healing Mushrooms* by Tero Isokauppila

Poisonous Mushrooms
- *Common Poisonous Plants and Mushrooms of North America* by Nancy Turner and Patrick von Aderkas

Growing Mushrooms
- *Growing Gourmet and Medicinal Mushrooms* by Paul Stamets
- *The Mushroom Cultivator: A Practical Guide to Growing Mushrooms at Home* by Paul Stamets and J. S. Chilton
- *DIY Mushroom Cultivation* by Willoughby Arevalo

Children's Books
- *Mike O'Rhiza* by Dr. Mike Amaranthus and illustrated by Linda Woodrow-Gray
- *The Mushroom Fanclub* by Elise Gravel
- *Fantastic Fungi Coloring Book* illustrated by Roman Daniel Eason

Websites
- Mykoweb.com
- Mushroomexpert.com
- Namyco.org
- Fantasticfungi.com
- Fungi.com
- Mycorrhizae.com

INDEX OF MUSHROOM COMMON NAMES

admirable bolete, 209, 227–228
American matsutake, 127–132
Artist's conk reishi, 43, 299–300
bearded milk cap (woolly milk cap), 72, 81–82
bear's head, 190, 195, 201–203
bitter bolete, 213, 216, 219, 221–222, 225, 228
black trumpet, 50–51, 61–63
bleeding milk cap, 72–73, 75–76
cauliflower mushroom, 285, 291–292
chaga, 39, 41–42, 44, 286, 303–304
comb hericium, 190, 195, 201, 203
deadly galerina, 114, 187, 189
death cap, 92, 95–96, 107–109, 114, 139, 187, 189
delicious milk cap, 72, 77, 79, 83
destroying angel, 92, 97, 99, 107–109, 114, 130, 139, 187, 189
false chanterelle, 50–51, 65–66
false morel, 234, 242–243, 257, 259, 261
false parasol, 115, 123–124
fire morel, 241, 247, 249
fly agaric, 17, 92–93, 101–103, 105–106, 110–112, 114
French black truffle, 267–268, 271, 277–279
gold caps (magic mushroom), 150, 153, 155–157, 159, 165
hedgehog, 20, 35, 60, 190–191, 193, 203, 206
hen of the woods, 17, 39, 44, 199, 285, 287, 289
honey mushroom, 85–90
liberty cap, 157, 159, 189
lion's mane, 17, 39, 44, 177, 190, 195–199, 201, 203, 207
manzanita bolete, 209, 223, 225
mountain blond morel, 25, 237, 252, 255
natural black morel, 245–247
Oregon white truffles, 268, 273, 277, 284
oyster mushroom, 39, 44, 179, 181–186
Pacific golden chanterelle, 18, 51, 53–55, 57, 66
panther amanita, 17, 92, 105–107, 114
parasol, 115, 117–119, 123–125
porcini, 25, 27, 32, 208–209, 211–213, 216, 221, 223, 227, 229–233, 282
prince, 137–142

red-belted conk, 286, 307–308, 318
red-pored bolete, 213, 216–217, 219, 222, 225, 228
Satan's bolete, 213, 215–216, 219, 222, 225, 228
shaggy mane, 124, 143, 146–149
shaggy parasol, 115, 119, 121, 124
Stuntz's blue legs, 163, 165
thimble morel, 257–258
turkey tail, 39, 44, 184–185, 286, 293, 295, 304
varnish conk reishi, 286, 296, 299
wavy cap, 66, 161–162, 165
western elfin saddle, 263–264
white chanterelle, 50–51, 54–55, 57, 59, 66
yellow foot chanterelle, 50–51, 59–60, 81, 193
yellow morel, 238, 251–252

INDEX OF SCIENTIFIC NAMES

Agaricus augustus, 137, 141–142
Amanita muscaria, 18, 101–102, 106, 111–113, 176
Amanita ocreata, 24, 35, 97, 108
Amanita pantherina, 105–106
Amanita phalloides, 35, 95, 108
Armillaria mellea, 85, 88–89
Aureoboletus mirabilis, 227
Boletus edulis, 208, 211–212, 230
Caloboletus rubripes, 221
Cantharellus formosus, 53–54
Cantharellus subalbidus, 54–55
Chlorophyllum molybdites, 123
Chlorophyllum rhacodes, 119
Coprinus comatus, 143
Craterellus cornucopioides, 61
Craterellus tubaeformis, 59
Fomitopsis pinicola, 307
Galerina marginata, 152, 163, 187, 189
Ganoderma applanatum, 299
Ganoderma oregonense, 296
Grifola frondosa, 287, 289
Gyromitra esculenta, 35, 243, 259, 261
Hericium abietis, 201
Hericium coralloides, 203
Hericium erinaceus, 195
Helvella vespertina, 263
Hydnum repandum, 191
Inonotus obliquus, 303
Lactarius deliciosus, 77
Lactarius rubrilacteus, 72, 75
Lactarius torminosus, 81
Leccinum manzanitae, 223, 225
Macrolepiota procera, 117
Morchella elata, 245–247
Morchella esculenta, 251
Morchella rufobrunnea, 253
Pleurotus ostreatus, 179
Psilocybe cubensis, 150, 153, 156

Psilocybe cyanescens, 161
Psilocybe semilanceata, 157, 189
Psilocybe stuntzii, 163, 171–173, 189
Rubroboletus pulcherrimus, 217
Rubroboletus satanas, 215
Sparassis radicata, 291
Trametes versicolor, 293
Tricholoma murrillianum, 127, 129
Tuber gibbosum, 267, 273–274
Tuber melanosporum, 267, 277
Tuber oregonense, 267, 273–274
Turbinellus floccosus, 65–66
Verpa bohemica, 243, 257

INDEX OF MUSHROOM ESSAYS, RECIPES, HISTORY

Amanita
Amanita muscaria and Christmas Traditions?, 111–113
Amanita Toxins, 114
Ancient Hallucinogen, 110
Murder, 108–109

American Matsutake
Fun Facts about Matsutake, 130–131
The Japanese American Matsutake Experience, 132–135

Boletes
My First Mushroom Hunt: Porcini, 229–230
Porcini Culinary Uses, 231
Recipe: Porcini Mushroom with Grilled Cheese, 232
Recipe: Simple Porcini Recipe, 233

Chanterelles
A Mushroom Trilogy, 68–70
Caring for and Preparing Chanterelles, 67
Recipe: Chanterelle Soup with Red Wine and Baguette, 71

Corals and Conks
How to Make Tinctures, 304–305
Reishi: A Rich Medicinal History and Use, 300–301

Hedgehog
Recipe: Sautéed Hedgehog, 206

Honey Mushroom
Recipe: Sautéed Honey Mushroom Caps, 90
The Ancient Humongous Fungus among Us, 88–89

Lion's Mane
Lion's Mane Medicinal Properties, 197–198
Recipe: Lion's Mane "Crab" Cakes, 207

Milk Caps
Recipe: Marinated Milk Caps, 83

Morels
Morels: What, Where, and When?, 234–237
Recipe: Morel Gnocchi with Peas and Prosciutto, 265–266

Oyster Mushroom
Recipe: Coconut Oyster Mushrooms with Ginger and Scallions, 186
Versatile Fungus, 184–185

Parasols
Recipe: Breaded Parasol, 125

Prince
A Prince of a Story, 140–141
Asparagus *Agaricus augustus*, 142

Psilocybe
Mushrooms Lessons Learned, 166–169
Psilocybin Mushrooms—A Reflection, 170
The Stoned Ape Hypothesis, 174–178
The Stutzii Secret, 171–173

Shaggy Mane
Fun Facts about Shaggy Manes, 147–148
Shaggy Mane–Crusted Parmigiana, 149

Truffles
A Very Brief, "Underground" History of the Truffle, 270–271
Hazel Hill: Black Truffle Plantation Extraordinaire, 279–281
Italian Truffles Are Pretty, Pretty Good, 282–283
The Amazing Fungal and Animal Connection, 313–314
Truffles, Trees, and Critters, 284

PHOTO CREDITS

All photographs are by the author except for the following:
Eileen Amaranthus, pp. 10, 105, 146, 147, 198, 239, 255
Dave Arora, pp. 249
Eric Ballinger, pp. 132, 133, 135
Dr. Efren Cazares, pp. 309, 310
Dr. Bruce Damer, pp. 176
Marcus Durban, pp. 25, 234
Dr. Todd Elliot, pp. 311 (bottom), 312 (top)
Joel Frank, pp. 182 (bottom)
Dr. Megan Frost, pp. 166
Tim Giraudier, pp. 6, 18 (top and bottom), 52, 140, 204, 286
Trish Glose, pp. 202
Linda Woodrow-Gray, pp. 35, 39, 93, 111, 138, 197, 326–329
Grant Jantzer, pp. 43 (bottom)
Louie Jeandin, pp. 5 (bottom left), 34, 43, 51 (top), 130, 194, 200, 231, 240, 241 (right), 290
Dr. Pam Kryskow, pp. 170
Gordon Longhurst, pp. 69, 70, 295
Dr. Dan Luoma, pp. 76
Dr. Nick Malajczuk, pp. 267, 268, 269 (top and bottom), 271, 279, 280
Breanne Pratt, pp. 196
David Steinfeld, pp. 5 (top), 57, 171, 296
Rafael Santini, pp. 42, 302
Dr. Jim Trappe, pp. 284, 313 (upper and lower)
Dustin Welch, pp. 15 (bottom), 28, 129

ABOUT THE AUTHOR

Dr. Mike Amaranthus believes *fungi* should emphasize *fun*. As a two-time cancer survivor, getting to know this amazing group of mushrooms has enriched and perhaps saved his life. He believes that a book can not only be accurate and have substance but also delight the reader with stories, history, possibilities, recipes, and much, much more.

Dr. Mike is a retired research soil scientist for the US Department of Agriculture and associate adjunct professor at Oregon State University. Dr. Mike received the USDA's highest Honor Award for scientific achievement after his twenty years of contributions as a USDA research scientist, authoring and coauthoring over one hundred scientific papers. Dr. Mike has several mushroom and truffle species named after him in his honor. Yes, having fungi named after you is an honor, not a curse!

Two decades ago, Dr. Mike and his wife, Eileen, founded Mycorrhizal Applications Inc., a biotechnology company focusing on growing and using mycorrhizal fungi to increase food production using fewer chemical fertilizers and pesticides. The globally groundbreaking products are now used in over thirty countries. Dr. Mike Amaranthus holds an undergraduate degree from the University of California–Berkeley and a PhD from Oregon State University.

Dr. Mike is active as a mentor and volunteer for youth and cancer support network organizations all over Oregon. Dr. Mike and Eileen are thankful for a full and warm home, raising five children and having eight grandchildren.

FRY, THRIVE, OR DIE
KEY TO MUSHROOMS WITHOUT GILLS

- The cap underside has a layer pores that are closely packed tubes.
- If the mushroom has gills, go to the key on PP.328-PP.329.
  - The underside of cap lacks a layer of tubes or pores.

If the mushroom has teeth or downward pointing spines with or without a well defined cap.

TEETH FUNGI (PP.190-PP.207)

If the mushroom lacks downward pointing spines.

If the mushroom is honeycombed with pits and ridges, saddle or brain shaped and has a stalk.

MORELS AND FALSE MORELS (PP.234-PP.266)

If the mushroom is a different shape or lacks a stalk.

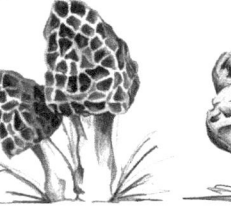

If the mushroom is rounded potato shape and fruits underground with a veined, marbled interior.

TRUFFLES (PP.267-PP.284)

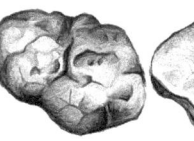

The mushroom has a fleshy cap and a central stalk that are soft and sponge-like.

BOLETES (PP. 208-PP.233)

The mushroom is woody, tough, bracket like, or coral-liked (branched) without a well defined cap.

CORALS AND CONKS (PP.285-PP.308)

To make a positive identification, *all* of the key features of your mushroom must be present as described for that species and, if they are not, you must assume it is not the correct mushroom. Its best to gather several specimens so you can look at the range of characteristics as the mushroom ages.

FRY, THRIVE, OR DIE
KEY TO MUSHROOMS WITH GILLS

→ If mushroom has gills or radiating blades (sometimes folds) beneath the cap.

→ If the mushroom does NOT have gills, go to the key on PP.326-PP.327.

↓ If the mature gills are free from the stalk, are white or yellow and the mushroom has a volva (sack, collar or scaly rings) and patches, warts or flakes.

AMANITA'S (PP.92-PP.114)

→ If not as above.

→ If the young gills are covered by a veil that forms a ring on the stalk.

→ If no veil covers the young gills and no ring forms on the stalk.

MILK CAPS (PP.72-PP.83)

→ If mushroom stalk snaps cleanly like a piece of chalk and gills when broken bleed a juice or latex.

→ If the stalk shows fibers, does not snap clearly and the gills when broken do not bleed juice or latex.

→ If the spores are white, yellow or pinkish.

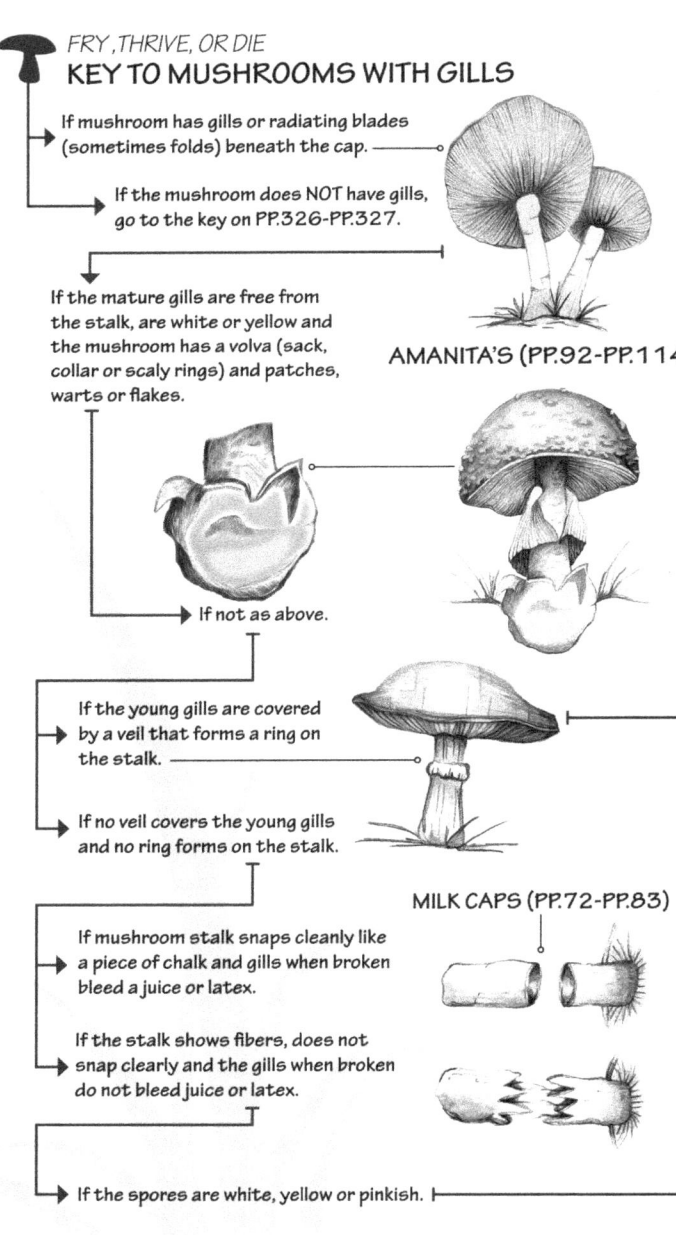

If the spores are chocolate brown to dark chocolate brown when mature, but pinkish when younger and gills free from stalk.

AGARICUS (PP.136-PP.142)

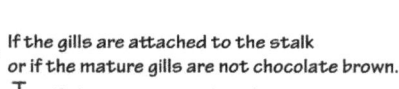

If the gills are attached to the stalk or if the mature gills are not chocolate brown.

If the spores are white (or greenish in one case) and the mature gills are white or yellow.

Light spored gilled mushrooms with a ring.

HONEY MUSHROOM, MATSUTAKE, PARASOLS (PP.84-PP.90, PP.126-PP.135, PP.115-125)

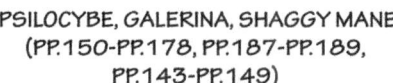

PSILOCYBE, GALERINA, SHAGGY MANE (PP.150-PP.178, PP.187-PP.189, PP.143-PP.149)

If the spores are rusty orange, brown, or black and the mature gills are not white.

Dark spored gilled mushrooms with a ring.

If the "gills" are blunt or thick and forked or cross veined and run down the stalk.

CHANTERELLES AND FALSE CHANTERELLES (PP.50- PP.71)

If the gills are blade like and run down the stalk and the cap and stalk is off center.

OYSTER MUSHROOM (PP.179-186)

CPSIA information can be obtained
at www.ICGtesting.com
Printed in the USA
JSHW040844210323
39212JS00004B/8

DUMB CRIMINALS & OVEREAGER COPS

A Paralegal's Guide to Criminal Law and Procedure

ZACHARIAH B. PARRY, J.D.

Las Vegas, Nevada

Dumb Criminals and Overeager Cops: A Paralegal's Guide to Criminal Law and Procedure
Copyright © 2018 by Zachariah B. Parry, JD.

All rights reserved. Printed in the United States of America. No part of this book may be used or reproduced in any manner whatsoever without written permission except in the case of brief quotations embodied in critical articles or reviews. This book is a summary or amalgam of substantive and procedural laws across the country and does not represent the laws in any given jurisdiction. It is not intended to substitute the advice of a licensed attorney. It is for instructional purposes for paraprofessionals in the legal field.

There are a number of references to real life cases, crimes, and popular culture. Many crimes, by their nature, are terrible and offensive. Although this book strives to be accurate, it in no way attempts to be terrible or offensive.

For information or to provide feedback, contact the author:

Dueling Peacock, Ltd.
880 Seven Hills Drive, Suite 210
Henderson, Nevada 89052
zach@paralegaltrainer.com

Book and Cover design by Zachariah B. Parry
ISBN: 978-1-717901-60-6

First Edition: July 2018

CONTENTS

PREFACE i
    DUMB AND OVEREAGER i

PART 1: CRIMINAL LAW 1

CHAPTER ONE 1
    INTRODUCTION TO CRIMINAL LAW 1
        How Laws Are Created 3
        Categories of the Law 8
        Theories of Punishment 11
        Classification of Crimes 13

CHAPTER TWO 16
    ESSENTIAL ELEMENTS OF A CRIME 16
        Physical Act 17
        Mental State 20
        Concurrence 23
        Causation 24
        Harm 26

CHAPTER THREE 27

ACCOMPLICE LIABILITY 27
Parties to a Crime 27
Scope of Liability 30

CHAPTER FOUR 33

INCHOATE OFFENSES 33
Attempt 34
Solicitation 38
Conspiracy 39

CHAPTER FIVE 43

CRIMES AGAINST THE PERSON 43
Assault and Battery 45
Mayhem 48
Homicide 50
False Imprisonment 54
Kidnapping 57

CHAPTER SIX 59

SEX OFFENSES 59
Rape 59
Statutory Rape 63

CHAPTER SEVEN 66

PROPERTY OFFENSES 66
Larceny 67
Embezzlement 68
False Pretenses 70
Robbery 72
Extortion 73
Receipt of Stolen Property 75
Forgery and uttering false instruments 77

CHAPTER EIGHT ... 80

OFFENSES AGAINST THE HABITATION ... 80
Burglary ... 81
Arson ... 85

CHAPTER NINE ... 87

OFFENSES INVOLVING JUDICIAL PROCEDURE ... 87
Perjury ... 88
Subornation of Perjury ... 90
Bribery ... 91

PART 2: CRIMINAL PROCEDURE ... 93

CHAPTER TEN ... 95

INTRODUCTION TO CRIMINAL PROCEDURE ... 95

CHAPTER ELEVEN ... 97

THE FOURTH AMENDMENT ... 97
Searches ... 99
Seizures, Arrests, and Detention ... 104
Warrant Requirements ... 105
Exceptions to the Warrant Requirement ... 112

CHAPTER TWELVE ... 121

THE EXCLUSIONARY RULE ... 121
Limitations ... 125

CHAPTER THIRTEEN ... 127

INTERROGATION AND CONFESSIONS ... 127
Fifth Amendment and Miranda ... 128
Fourteenth Amendment ... 133
Sixth Amendment ... 135

CHAPTER FOURTEEN ... 138

PRETRIAL PROCEDURES ... 138
Preliminary Hearing ... 141

Indictments and Grand Juries..143
Arraignment..146
Speedy Trial..147
Discovery...148

CHAPTER FIFTEEN 152

TRIAL 152

Right to Public Trial...153
Right to a Jury Trial...157
Right to Counsel...159
Right to Confront Witnesses...159
Burden of Proof..163

CHAPTER SIXTEEN 168

SENTENCING AND PUNISHMENT 168

Sentencing Rights...169

INDEX 174

PREFACE

Dumb and Overeager

WHEN I STARTED LAW SCHOOL, I really had no idea what to expect. As far as I knew, going back all the way to Adam and Eve, I was the first attorney in my direct line.

Before class even started my first semester of law school, as soon as I picked up my textbooks, I felt like I was in over my head. They all had different authors, but still looked like they belonged in a set—they were thick, hardbound, had gold lettering, and were either red on blue or black on red. And they all had super creative names, like "Business Law: Text and Cases," or "Fundamentals of Criminal Law." At $200 per book, I felt like I deserved a better title.

As I flipped randomly through a few of the books and saw that they were filled with the entire text of actual legal cases, with a smattering of commentary, I began to wonder if I had made a mistake deciding to become a lawyer. I could not imagine myself being able to make it through all the books without going crazy from boredom.

Turns out the law is actually quite interesting. It doesn't need to be dressed up at all for it to be a captivating read. Even something that sounds (at least to my untrained ears) as droll as "property law," I learned some fascinating legal principles and examples, including how a case from England about a foxhunt helped determine who the proper owner would be when two people fought over Mark McGuire's record-breaking home run ball.

The problem with legal textbooks is not so much that they aren't covering interesting topics. It's that they are covering very interesting topics in a very uninteresting way. It's like numbing your tongue with Anbesol before eating a hamburger with an Applewood-smoked-bacon-and-cheddar-cheese infused patty. If you're going to numb your taster, you might as well just heat up some frozen hamburger and call it a meal.

So it is with most paralegal textbooks. They are informative, yes. But conveying information should not be the only goal of teaching. Being truly successful in the legal field—whether as a judge, lawyer, or paralegal—is about much more than just cramming your head full of useful (and some not-so-useful) information and then regurgitating it. It's about finding your passion and loving what you're doing. And trust me, it's not hard to become passionate about the law.

After all, the law is very real. It impacts all of us. Sometimes it shields us. And sometimes it hurts us. It might be the only way we can get justice. But it also might feel unfair. It has been around forever. But it also changes constantly.

I wanted this textbook to not only convey the information that will form the foundational backdrop of your paralegal career, but I also wanted you to enjoy reading it. I wanted you to look forward to reading it. Maybe even read ahead. Or share excerpts with your family and friends.

Because if you're doing that, you're learning to love the law. You are discovering that all around you, every day, are examples of legal principles coming to life—often in colorful and interesting ways.

I have researched far and wide for examples of the principles being taught in this book. I have also delved deep into my not-inconsiderable store of pop culture references. My goal was that for every major legal principle taught in this book, I would either give you an analysis using some pop culture reference, a famous case example, something that would fall under the "dumb criminals and overeager cops" category, or some combination of the three.

The latter are perhaps the most fun. Those are the stories you just can't make up. And as I read countless examples of idiot criminals, I inadvertently stumbled upon the title of this book. Or at least the original title, which was "Criminals Are Dumb." When I shared the prototype cover with some old friends, one of them, a criminal defense attorney in New Mexico, Jamie Askar, called me out on an insensitive title that only perpetuated harmful stereotypes. I appreciated his perspective, and with some help from him, we retitled the book not only to be more accurate, but more evenhanded.

See, the title is not just an awkward attempt to be clever. It's rather accurate. You may have heard the saying that laws are written in blood. It's true. We have laws on the books because something bad happened, we knew it was wrong, but we didn't have the rules in place to criminalize it. So, we made it illegal. Thus, our criminal laws—the ones that apply to all of us—were developed to respond to terrible decisions made by members of our community.

The same is true of procedural law—the rules designed to protect the accused against overly ambitious police officers and lawyers who are so focused on putting the "bad guy" away that they forget that the "bad guy" is innocent until proven otherwise. This requires that the rules be followed to make sure the accused gets a fair opportunity to mount a defense.

As you read this book, I hope you not only learn the fundamentals of criminal law and procedure (which is what this book would be called if I tried to market it to a law school), but you also get a glimpse of how fascinating and rewarding it can be to jump into the trenches of the veritable battlefield we call the law.

Happy learning.

~Zach Parry

PART 1: CRIMINAL LAW

CHAPTER ONE

Introduction to Criminal Law

AS LONG AS THE HUMAN race has been writing stuff down, we've been creating rules. And as long as we've had rules, we've been breaking them.

The oldest known criminal code in existence is the Code of Ur-Nammu Sumer (in what is modern-day Iraq) from the 22nd Century BC. Among its edicts are proscriptions that range from the familiar ("If a man commits murder, that man must be killed") to the improbable ("If a man, in the course of a scuffle, smashed the limb of another man with a club, he shall pay one mina of silver"[1]) to the bizarre ("If a man is accused of sorcery he must undergo ordeal by water; if he is proven innocent, his accuser must pay 3 shekels[2]) to just plain wrong ("If a man proceeded by force, and deflowered the virgin

---

[1] One mina is about 498 grams or 17.5 oz. Although it's unfair to compare the value of silver four millennia ago to the value of silver today, that comes to $287.88 in March 2018 silver prices.

[2] A shekel was a unit of weight and was the equivalent of 8.3 grams. Three shekels of silver would be 24.9 grams or $13.20 in March 2018 silver prices.

female slave of another man, that man must pay five shekels of silver"[3]).

More familiar to modern readers will be the Code of Hammurabi from Babylon (18th Century BC Iraq) and the Ten Commandments handed down by God to Moses in Egypt (approx. 13th Century BC), both of which established societal standards of conduct.

Among the very first federal laws enacted by an infant United States Congress was *An Act for the Punishment of Certain Crimes Against the United States* (1790), which, among other things, established a punishment for treason, piracy, and counterfeiting.

Those guilty of treason (adhering to enemies of the U.S. and giving them aid and comfort) were sentenced to death. Piracy (piratically and feloniously running away with a ship or vessel) was also a capital offense. The First Congress stretched the limits of its collective creativity in establishing a punishment for counterfeiting (falsely making, altering, forging or counterfeiting any certificate, indent, or other public security). You guessed it, those adjudged guilty of counterfeiting forfeited their lives.

As the United States has matured, many of the draconian penalties levied in the days of yore are not so severe or black and white.[4]

Treason, for example, codified into modern federal statute in 1948, allows for punishment ranging from imprisonment for at least 5 years with a fine of at least $10,000 on the low end to death on the other end of the spectrum.[5] And although treason has been a crime since before we

---

[3] $22.00.
[4] For a discussion of the evolution of sentencing in the United States, turn to Chapter 16.
[5] 18 U.S.C. § 2381.

declared our independence from Britain, no one in the United States' 242-year history has ever been executed for treason against the federal government.[6]

The punishment for counterfeiting and piracy have also grown less severe. Counterfeiting carries a maximum penalty of 20 years' imprisonment,[7] and piracy has nothing to do with seafaring vessels anymore but applies when kids living in their parents' basements down-load media without paying for it.[8] You can't watch a movie at home without the Federal Bureau of Investigation reminding you[9] that unauthorized copying is subject to a penalty of up to 5 years and $250,000.[10]

HOW LAWS ARE CREATED

Anyone who has been through high school should have at least a rudimentary understanding of how a law is born.

Anyone can draft a **bill** (a proposed new law), but a member of either chamber of Congress (either the House of Representatives or the Senate) has to sponsor it. The bill then

---

[6] Larson, Carlton F.W., "Five myths about treason." *The Washington Post*, February 17, 2017 (last accessed March 30, 2018).
[7] 18 U.S.C. § 471.
[8] There is still a code on the books dealing with sea piracy, the penalty for which is life imprisonment. 18 U.S.C. § 1651.
[9] I wonder if that FBI warning is copyrighted. If so, I wonder if "ironic use" is a defense to copyright infringement.
[10] 17 U.S.C. §§ 501; 506.

gets referred to a committee for a preliminary determination of the bill's chances for passage. If it makes it through the committee process (which could involve subcommittee hearings, votes, and amendments), it is referred to the other chamber where it goes through a similar process and either gets approved, rejected, ignored, or modified and sent back to the originating chamber.

If both chambers approve the bill without changing it, the President has ten days to **veto** the bill (overrule it). If the President vetoes the bill, Congress has an opportunity to override the veto with a supermajority vote (two-thirds). If the President does not veto the bill, or his veto is overridden, the bill becomes law.

Sometimes Congress acts in a unified manner and the process is relatively quick. Other times the process is fraught with political tension and back-room deals. Consider two examples: The USA PATRIOT Act and the Affordable Care Act (Obamacare).

The USA PATRIOT Act

After the 9/11 terrorist attacks, our nation recoiled. Taking the initiative in the immediate aftermath, the Bush Administration proposed legislation designed to combat terrorism. They presented their proposal to Congress less than two weeks after the attack, on September 24, 2001.

It made it past committee and to the floors of both chambers of Congress for a general vote on October 11, 2001, though the two different versions of the bill were slightly different. The Senate passed the bill with a vote of 96-1. The House was unanimous in its approval.

> "All bad precedents begin with justifiable measures."
> ~Julius Caesar

On October 24, 2001, the House proposed and

unanimously passed a new bill that reconciled the differences between the two bills. This new bill was called the "Uniting and Strengthening America by Providing Appropriate Tools Required to Intercept and Obstruct Terrorism Act," the U.S.A. P.A.T.R.I.O.T. Act, or USA PATRIOT Act.

The Senate voted again the next day, this time 98-1 in favor, and President George W. Bush signed the bill into law on October 26, 2001.

Thus, just 45 days after the largest terrorist attack in our history, the United States in one sweeping gesture broadened its police and surveillance powers. They probably could have done it even faster, but most of that time was spent trying to come up with a way to turn the word "Patriot" into an acronym.

The Affordable Care Act

House Democrats proposed a healthcare overhaul in July 2009. At the time, Democrats in the Senate had a 60-seat supermajority, which would have made passage of the act possible without any support from Republicans. That changed when Massachusetts senator Ted Kennedy—a Democrat—passed away in August 2009.

In November, the first vote was cast in the House. The score: 220–215 in favor (with only one Republican voting in favor and 39 Democrats voting against).

The Senate voted on their version Christmas Eve: 60 in favor and 39 against, with one senator not voting.

Different versions of the bill had now passed in each chamber, but in January 2010, running on a platform against the new healthcare law, Republican Scott Brown won the special election in Massachusetts to take over Ted Kennedy's spot. House Democrats change their tactics, knowing the

Senate could not get the 60 votes needed to overcome the Republican filibuster.

The House decided instead to pass the version that had already been approved in the Senate—which would not require another Senate vote, which they would not win. Instead, they created a new bill that had promises that the Affordable Care Act would be amended in the future to meet the demands of the House. Because that bill was merely budgetary, it was not subject to a Republican filibuster (per the Congressional Budget Act of 1974), which meant it could pass in the Senate with only 51 votes.

That satisfied most House Democrats, but there were some pro-life Democrats who were hesitant to vote for the Senate version of the bill, which included federal funding for abortion. Since that issue could not be addressed in the budgetary bill, to get their votes, President Obama issued an **executive order** (a President-issued directive that has the force of law) that reaffirmed a commitment not to use federal money to fund abortion even under the Affordable Care Act.

With enough representatives in the House satisfied, they voted again on the bill that had already passed in the Senate, and it passed 219–212. President Obama signed it into law March 23, 2010—almost nine months after it was proposed.

Criticism of both the PATRIOT Act and Obamacare abound—opponents complain that the PATRIOT Act was rushed and passed only because of the high emotions resulting from 9/11 and that Obamacare represented the worst of politics—one political party steamrolling the other and stopping at nothing to introduce legislation that no one had actually read and that the country neither needed nor wanted.

But that's exactly how laws are made. We elect the

representatives, and they create the laws that they believe are in our best interest. If as a populace we disagree with them, we vote them out and someone else in. In the meantime, we have to follow the laws they establish or live with the consequences of breaking them.

Not all federal laws are as controversial and publicly known as the PATRIOT Act and Obamacare. In fact, even after giving government twenty-four decades to figure out a fair and just criminal system, there are still some bizarre criminal laws on the books.

Among the approximately 5,000 criminal statutes and over 400,000 regulations criminalizing certain behavior[11] are the following, which are still on the books and enforceable:

Crime[12]	Statute/Regulation
To sell "Turkey Ham" as "Ham Turkey" or with the words "Turkey" and "Ham" in different fonts.	21 USC § 461 & 9 CFR § 381.171(d).
To wash a fish at a faucet if it's not a fish-washing faucet, in a national forest.	16 USC § 551 & 36 CFR § 261.16(c).
To let your pet make a noise that scares the wildlife in a national park.	18 U.S.C. § 1865 & 36 C.F.R. § 2.15(a)(4).
To injure a government-owned lamp.	40 U.S.C. § 8103(b)(4).
To sell onion rings resembling normal onion rings, but made from diced onion, without saying so.	21 USC § 333 & 21 CFR § 102.39.
To ride a moped into Fort Stewart without wearing long trousers.	8 U.S.C. § 1382 & 32 CFR § 636.28(g)(iv).
To hunt doves or pigeons with a stupefying substance.	16 U.S.C. § 707; 50 CFR §§ 20.21(a) & 20.11(a).
To say something so annoying to someone	7 U.S.C. §1011(f) & 36 CFR §

---

[11] Jason Pye, "19 Ridiculous Federal Criminal Laws and Regulations," *Freedom Works*, January 14, 2016, available at http://www.freedomworks.org/content/19-ridiculous-federal-criminal-laws-and-regulations (last accessed March 30, 2018).

[12] If you are interested in learning more about laws like these (to make sure you don't accidentally commit a crime), subscribe to @CrimeADay on Twitter, which is the handle of the organization that compiled this list.

Crime[12]	Statute/Regulation
that it makes them hit you in a national forest.	261.4(b).
To attempt to change the weather without telling the Secretary of Commerce.	15 U.S.C. §§ 330(a) & (d).

On an unrelated note, members of Congress—the ones who make these laws—have about a 15% approval rating.[13]

CATEGORIES OF THE LAW

Broadly speaking, most laws can be categorized as belonging to one of two families: criminal law and civil law.

> "Law is the embodiment of the moral sentiment of the people."
> ~William Blackstone

Criminal v. Civil Law

Civil law is the system of laws that sets enforceable boundaries in the relationships between individuals (or businesses). **Criminal law**, on the other hand, is a set of behavioral rules that the government sets, a violation of which subjects the violator to penalties.

Consider these differences between civil and criminal law:

	Civil Law	Criminal Law
Who is wronged	The individual	Society
Who enforces the law	Wronged individual (plaintiff)	The government (police, prosecutor)
How the law is enforced	Lawsuit	Arrest and prosecution
What is sought to be proven	Liability	Guilt
Burden of proof	Usually preponderance of the evidence	Beyond a reasonable doubt

---

[13] "Congress and the Public," Gallop February 2018, available at http://news.gallup.com/poll/1600/congress-public.aspx (last accessed March 31, 2018).

	Civil Law	Criminal Law
Consequence	Usually money damages	Loss of money (fine), liberty (imprisonment), life (death)

The remedies for civil and criminal violations of the law are not mutually exclusive. Thus, the same act may constitute both a civil wrong (called a tort) and a crime.

Remember O.J. Simpson? He was prosecuted for murder and **acquitted** (found not guilty) in 1995. The jury was not convinced, beyond a reasonable doubt, that he had murdered Nicole Simpson and Ron Goldman.

Not satisfied with the verdict in the criminal trial, the families of the victims sued O.J. in civil court and were able to convince a jury by the preponderance of the evidence that O.J. was liable for the deaths. The jury awarded $8.5 million in compensatory damages and $25 million in punitive damages.

The lower standard of proof was not the only advantage the plaintiffs in the civil case had over the prosecutors in the criminal case. Because the consequences of a criminal conviction are typically more severe than a finding of civil liability, there are additional protections afforded the defendant in a criminal case.

For example, O.J. had the right to refuse to testify on his own behalf in the criminal case (called the **right against self-incrimination**[14]), but in the civil case, he was subject to the plaintiffs' trial subpoena and had to take the stand. Thus, the jury was able to hear from O.J.'s own lips what his defenses were and judge his credibility for themselves. They apparently had reasons to doubt the sincerity of his claim of innocence. And that was even before he wrote *If I Did It:*

---

[14] For a fuller discussion on this topic, turn to 121.

Confessions of the Killer.[15]

Administrative Crimes

Several agencies operate within the confines of state and federal governments. The Environmental Protection Agency, Food and Drug Administration, and Department of Homeland Security are among the multitude of federal agencies.

These agencies promulgate administrative rules and regulations that, while not criminal laws themselves, are nonetheless punishable if violated. The Environmental Protection Agency publishes on its website a "Summary of Criminal Prosecutions" resulting from violations of environmental regulations.[16]

You may recall in 1984 when Walter Peck,[17] an inspector for the EPA in New York, expressed concerns about the potential negative environmental impact of a privately owned and operated storage facility in Lower Manhattan. He suspected the presence of "noxious, possibly hazardous waste chemicals."

After securing a court order authorizing his entry and inspection of the containment unit, Walter Peck threatened the owners for at least half-a-dozen environmental violations before he shut down the power grid, causing a massive explosion that released hundreds of ghosts into New York City.

The Ghostbusters were powerless to stop the EPA, but eventually saved the city, including defeating the Stay Puft

---

[15] O.J. Simpson originally titled his book, "If I Did It," but after a judge awarded publishing rights of the book to Ron Goldman's father, Fred Goldman, to help satisfy the $33.5 million civil judgment, Goldman renamed the book "I Did It."

[16] Environmental Protection Agency, "Summary of Criminal Prosecutions," available at https://cfpub.epa.gov/compliance/criminal_prosecution/ (last accessed March 31, 2018).

[17] Ghostbusters, ©1984.

Marshmallow man, and with perfect poetic justice, inadvertently dumped gallons of liquefied marshmallow onto an unsuspecting but deserving Walter Peck.

THEORIES OF PUNISHMENT

Criminal law was created to protect society from wrongs that are offensive to the conscience (***malum in se***) and necessary to promote social order (***malum prohibitum***). Punishments represent society's moral condemnation of the behavior and to be effective, should promote certain underlying purposes. These stated purposes include incapacitation, deterrence, retribution, rehabilitation, and education.

Incapacitation occurs when we remove the criminal from society, taking away any further opportunities for violations, which in turn protects members of society. We succeed in incapacitating criminal offenders by arresting, restraining, and detaining them, which can be done temporarily (including probation and limited incarceration) or permanently (being sentenced for life or to death).

The Green River Killer, for example, who was convicted for the murder of 49 women and girls (and who perhaps killed many more) is currently serving 49 life sentences for the murders plus another 480 years for evidence tampering. Since his arrest in 2001, his incapacitation has prevented any further killings. He also has not been able to tamper with any more evidence.

Deterrence is the idea that when it is known that punishments are attached to certain behaviors, those behaviors are less likely to happen. **Specific deterrence** is punishment aimed at keeping the criminal from repeating the criminal behavior to avoid repeated punishment. **General deterrence** has a much broader goal. We hope that because there are consequences to criminal behavior—including punishment and the stigma of being a convicted felon—that society's members will avoid the behavior to avoid the consequences.

> "It may be true that the law cannot make a man love me, but it can stop him from lynching me, and I think that's pretty important."
>
> ~Martin Luther King, Jr.

> "Men are not hanged for stealing horses, but that horses may not be stolen."
>
> ~George Savile

Anyone who has ever been wronged understands the desire for **retribution**—punishing the wrong-doer out of a sense that it is right for a wrongdoer to suffer negative consequences for negative behavior. Retribution also gives the wrongdoer an opportunity to own up to the crimes by accepting the punishment and serves to deter vigilante justice—if society knows the government will mete out punishment, individuals are less likely to do so on their own. If the Gotham Police had been able to keep crime down on their own, Batman never would have been the hero Gotham needed.

Edmund Dantes—the Count of Monte Cristo—embodies well the idea of retribution. Although he was innocent, conspirators succeeded in convicting him as a political criminal, and he was imprisoned for fourteen years in the inhumane conditions at Chateau D'If. After his release, he acquired a massive fortune, declared himself a count, and one by one exacted revenge on each of the conspirators. Retribution is a theme we well understand.[18]

Perhaps more noble than retribution, **rehabilitation** is also a goal of punishment. Although there is considerable debate about the effectiveness of punishment as a rehabilitative tool, we still hope that by imposing punitive measures on our criminals, that they will take the opportunity to better themselves and acquire the desire to reincorporate themselves as productive members of society.

The final stated purpose of punishment is **education**. Because the criminal justice process is public, including trial and conviction, the process itself serves to educate the public about what is legally acceptable and what is not, and to promote a respect for the law.

CLASSIFICATION OF CRIMES

Criminal law has two components: substantive law and procedural law. **Substantive law** identifies crimes and their corresponding punishments. **Procedural law** establishes the rules that the rule enforcers have to follow.

When O.J. Simpson and his cohorts broke into room 1203 at the Palace Station in Las Vegas, Nevada to take sports memorabilia under threat of force, it would be substantive law that would tell you that these actions constitute burglary, robbery, conspiracy, and assault with a deadly weapon (among others).

Procedural law would tell you that to convict O.J., you have to charge him, arrest him, produce the evidence against him, and take the case to trial.

And if you're O.J., common sense should have told you

---

[18] Alexander Dumas, *The Count of Monte Cristo*, 1845.

after very publicly getting away with murder that you should never again place yourself in a position where a jury gets to decide your fate.[19]

Criminal Substantive Law: Malum in Se v. Malum Prohibitum

Substantive law can be further classified. *Malum in se* is a Latin term meaning "evil in itself" or "inherently wrong." It describes actions that are inherently wrong—like murder, rape, arson, and theft.

Malum prohibitum, or "wrong because prohibited," refers to those laws that criminalize behavior that is not in itself bad, but that is nonetheless outlawed for some socially desirable purpose, like driving on the left side of the road, gambling (in most of the country), and selling alcohol to minors.

Criminal Substantive Law: Classification of Crimes by Degree

To promote fair and equitable treatment of those committing crimes, and to avoid punishments that are too light or too harsh, we grade crimes by degree. Different jurisdictions are going to classify crimes differently, but as a general rule, the less severe the crime, the lighter the punishment.

A **misdemeanor** is a minor crime, and in most U.S. jurisdictions, carries with it a punishment of no more than one year of incarceration in a jail (as opposed to prison). A **felony** is a greater offense, which is punishable by more than one year of prison time.

---

[19] O.J. was convicted on all counts and sentenced to 9 to 33 years. He served 9 years before he was released on October 1, 2017.

A **capital offense** is the most severe class of crime, usually reserved for the most severe acts (like piracy in the 18th century) and is a felony punishable by death. Capital punishment — being sentenced to death—has been banned in 18 states and the District of Columbia.

> "I believe ... that while all human life is sacred there's nothing wrong with the death penalty if you can trust the legal system implicitly, and that no one but a moron would ever trust the legal system."
>
> ~Neil Gaiman

Pairing a crime with a punishment is much more complicated than calling it a misdemeanor or felony, though. Courts will consider **mitigating factors** (circumstances of the crime that make a lesser punishment fair) and **aggravating factors** (conditions that encourage the imposition of a more severe punishment.[20]

---

[20] Chapter 16 provides a more thorough discussion of sentencing.

CHAPTER TWO

Essential Elements of a Crime

TO CONVICT SOMEONE OF A CRIME, the prosecutor must be able to compare the behavior of the accused to some measurable standard. That means each crime must have a clear definition and well-defined elements. The **elements** of a crime are like a checklist—the components that must be proven to establish that the crime occurred.

Almost every crime has at least the following elements:

1. Physical act (*actus reus*);
2. Mental state (*mens rea*);
3. Concurrence;
4. Causation; and
5. Harm.

Legal professionals trying to prove the commission of a crime must prove these elements beyond a reasonable doubt.

Depending on the crime, the elements may take different

shapes, but they must all be present to meet the burden of proof. A paralegal will not be attempting to convince a jury that these elements exist, but it will nonetheless be essential for a paralegal in a criminal practice to be able to recognize the elements in a given set of facts for client screening purposes and as a driving force of discovery. This is true whether a paralegal works for the prosecution or defense.

PHYSICAL ACT

Not all lawyers are smart. Most of us are faking it to some degree. Throwing around Latin terms makes us feel smarter. Meet **actus reus**, a Latin phrase meaning "guilty act." Put simply, it means that (almost) every crime must have as one of its components a wrongful physical act (or omission) by the defendant.

Pulling the trigger of a firearm is the *actus reus* component of a gun-related murder. Entering someone's house to take a child is the *actus reus* of a kidnapping. Physically striking a person is the *actus reus* of a battery.

To constitute *actus reus*, there must be voluntary bodily movement. Voluntariness is a measure of the will. A muscle spasm, actions performed while unconscious, and other physical movements that were not the product of the actor's determination do not implicate the *actus reus* element of a crime.

Sometimes my kids play the "stop hitting yourself" game with each other.[21] It's not their favorite game, even if they can't help but laugh the whole time; but if hitting yourself were a crime, they wouldn't be guilty because the self-

---

[21] If you're not familiar with the game, Google "stop hitting yourself gif." You're welcome.

inflicted hits are not voluntary.

Sometimes the physical act can actually be the failure to act. If there is a legal duty to act, the actor is aware of the duty, and the duty is reasonably possible to perform, the failure to act creates an **omission** that rises to the level of *actus reus*.

Legal duties can arise from a statute, a contract, a special relationship, by voluntary assumption of care, and by putting someone else in peril. In those circumstances, the failure to act could constitute the "guilty act" or *actus reus* of a crime.

In the series finale of *Seinfeld*, Jerry Seinfeld and his friends were tried and convicted for failing to render aid to the victim of a carjacking they witnessed and filmed, in violation of a duty-to-rescue law in Latham, Massachusetts.[22]

The Three Amigos, contracted to protect the village of Santo Poco from the tyranny of the infamous El Guapo, could have been adjudged guilty of a crime when they turned their backs shortly after El Guapo and his henchmen arrived to terrorize the village folk and kidnap one of the women.[23] Their contractual duty to act is enough.

Sometimes the physical act is nothing more than mere **possession**. Like other "guilty acts," context is often key to determining whether possession of a particular item in a particular way qualifies. For example, possession of a firearm is less important than what one does with a firearm, but if it's concealed or carried by a felon that could be enough, even if

---

[22] In reality, there is no such law in Latham, and most of the duty-to-rescue laws on the books merely require the reporting of certain crimes (for example, California requires persons with knowledge of the rape of someone under 14 to report it and doctors must report gunshot wounds), and none of them require people to put themselves in danger to do it.

[23] *The Three Amigos*, ©1984. There is a happy ending to this story. The three cowardly amigos, Lucky Day, Dusty Bottoms, and Ned Nederlander, find their courage and confront the terrifying villain, rescue the damsel, help the villagers find their own courage, and together they defeat the horrible, evil, murdering, villainous monster, El Guapo.

the gun is never used.

Possession does not require literal physical possession, either. That's the difference between **active possession** and **constructive possession**.

It is enough that it is within the defendant's dominion and control. Illegal drugs in the trunk of a car are in the constructive possession of the driver even though he can't reach them while he's driving. Child pornography saved on a defendant's computer is in that defendant's constructive possession even if he doesn't have the computer with him.

Catching the Health-Conscious Drug Abuser

The Granite Shoals Police Department in Texas came up with a brilliant idea to combat intellectually challenged criminals' possession of illicit drugs. Either that or the department was trying to settle a debate about whether it's meth or heroin that has stronger brain-cell-killing potential. They posted on the department Facebook page a seemingly sincere public service announcement with a friendly warning: "If you have recently purchased meth or heroin in Central Texas, please take it to the local police or sheriff department so it can be screened with a special device. DO NOT use it until it has been properly checked for possible Ebola contamination!"

Twenty-nine-year-old Chastity Hopson saw the post, and since she has always been careful about not putting anything dangerous in her body, requested the complimentary Ebola-screen for her stash. The good news is that there was no Ebola in her drugs. The bad news is the resulting possession charge.

In other good news, those who picked meth over heroin earned $20 from the department pool.

In some narrow circumstances, a party can be liable for criminal conduct without engaging in any wrongful act (i.e.

without the *actus reus* requirement). This is true in **vicarious liability** cases, which applies when one party is liable for the criminal behavior of another—like where a corporation is liable for the criminal conduct of its employees.

MENTAL STATE

A physical act alone is not enough to criminalize a behavior. The same act in one circumstance may be wrong, and in another, it may be right. In most, but not all cases, there must also be a mental component. We call that **mens rea**, or "guilty mind."

Mens rea separates murder from justified killing, a kidnapping from a rescue, and a battery from self-defense. Consider the physical acts discussed as examples of *actus reus*: pulling a trigger, taking a child from someone's home, and striking a person.

> "Even a dog distinguishes between being stumbled over and being kicked."
> ~Oliver Wendell Holmes, Jr.

Soldiers who pull the trigger of a firearm with the intent to kill the enemy during a wartime battle are justified in engaging in the same physical act that in other circumstances would be murder. If the child in someone else's house is yours, the act of taking the child is no longer kidnapping, but a rescue, and if you physically strike someone who is attacking you, what would otherwise be a battery is now justified self-defense.

You can see that what someone is thinking when they perform a certain action is critical to determining whether their behavior is criminal. And because we have not yet developed mind-reading technology, to prove what someone's intent was, we have to look at the evidence surrounding the behavior. In the kidnapping/rescue example, it would be easy enough to prove who the child belongs to, for example.

The degree of intent is the measure of a guilty mind. **Specific intent** is a subjective desire or knowledge that the prohibited result will occur. Because the defendant will typically not admit his or her intent in committing a certain act (and the Fifth Amendment says they don't have to speak at all[24]), the attendant circumstances become crucial for crimes requiring proof of specific intent. That's why prosecutors look for motive. They help reveal the intent or purpose behind the crime.

Consider the specific intent required for each of these specific-intent crimes:

Specific Intent Crimes	
Crime	Intent that Needs to Be Proven
Assault	Intent to commit battery
Burglary	Intent to commit a felony in someone's home
Conspiracy	Intent to further an unlawful objective
Embezzlement	Intent to defraud
False Pretenses	Intent to defraud
Forgery	Intent to defraud
Larceny	Intent to deprive another of their property

General intent, on the other hand, is much broader. For general intent crimes, the prosecutor merely needs to prove that the defendant meant to do the act prohibited by law,

---

[24] *See* page 121 for a discussion on the Fifth Amendment privilege.

regardless of whether the results were intended. Rather than purposefully or intentionally committing harm, general intent crimes usually only require proof that it was done knowingly or voluntarily.

Battery is a crime of general intent. It is enough that the person swung a baseball bat knowing that it would, or at least was likely to, make physical contact with someone.

The measure of intent in murder and arson cases is **malice**. Legislators wanted to deprive those who commit these crimes from the defenses available to other specific intent crimes. The measure of malice is reckless disregard for an obvious substantial risk. Discharging a firearm into the air during Mardi Gras in New Orleans creates a high risk of death when the bullet returns to the ground. Someone who engages in this kind of behavior could be prosecuted and convicted of murder even if she had no specific designs to kill anyone.

Strict liability crimes are those that don't require a *mens rea* component (or may require it for some elements of the crime but not others). The act alone, even if done ignorantly or innocently, is enough. Statutory rape is a strict liability offense. The adult who engages in sexual relations with a minor below the age of consent will find no defense in claiming that she thought the minor was an adult (i.e., she did not think she was committing a crime and did not intend to commit a crime).

Speeding is a strict liability crime. You merely have to intend to be traveling at the rate you are traveling. You don't have to know what the speed limit is or intend to break the law to be guilty.

For those states that follow the Model Penal Code—a system of uniform laws introduced to provide a model that jurisdictions could adopt to promote uniformity among jurisdictions—there are four classifications of intent: purposely, knowingly, recklessly, and negligently.

An act is committed **purposely** when it is done with the intent to cause a certain result. This is similar to specific intent.

Doing something **knowingly**, like general intent, requires proof that certain conduct is very likely to cause a certain result.

Recklessness, comparable to malice, measures a gross deviation from the standard of care of an ordinary person or a conscious disregard for a high risk.

The final and lowest measure of intent is **negligence**, which is a measure of carelessness that creates an undue risk of harm.

CONCURRENCE

To prove a defendant guilty of a crime, a wrongful act and guilty mind (*actus reus* and *mens rea*) alone are not enough. The wrongful act must have been motivated by the guilty mind. This is called **concurrence**. It is also sometimes referred to as simultaneity or contemporaneity.

If the *actus reus* is not concurrent with *mens rea*, then no crime has been committed. Except for strict liability crimes, which do not require the element of *mens rea*.

Suppose that a man strikes a woman in the head with intent to kill her. She passes out and appears to be dead. Believing her dead, he throws her body into the river. But she is not dead and, unconscious in the river, she dies of drowning. But the intent to kill did not align with the act that killed her: when he struck her and meant to kill her, he did not. When he tossed her into a river and actually killed her, he did not have the intent to kill her, believing her already dead.

If there is no concurrence for the crime of murder, he cannot be convicted of murder. He may, however, be convicted of manslaughter, a lesser offense. [25]

In this example, there may be concurrence, however, in jurisdictions that recognize the single transaction principle: if the sequence of *actus reus* events can be said to be a single transaction (e.g., the knocking out and disposing of a person's body), and the *mens rea* occurs before or during the sequence. In these jurisdictions, the accused will find no safe harbor in a concurrence defense.

CAUSATION

The causation necessary to be proven in criminal cases is parallel to causation in civil cases. There are two components: actual cause and proximate cause.

An **actual cause** is a direct cause—something that would not have happened but for the crime. If it is true that "victim would not have died but for accused pulling the trigger," then it is also true that the accused pulling the trigger was an actual cause of victim's death.

There are, however, an infinite number of actual causes to any event, most of them too remote for courts and lawmakers to feel comfortable holding them accountable.

For example, suppose that the accused who pulled the trigger was the son of Bob and Mary. Bob and Mary meet forty years ago at a school dance because Bob's best friend Bill Gates introduced them. "But for Bill Gates introducing Bob to

---

[25] There is a similar case in England's *R v. Church*, 2 AER 72 (1965).

Mary, their son—the accused—would never have been born, would never have pulled the trigger, and the victim would still be alive." Because that is true, it is also true that Bill Gates' introduction of Bob to Mary is an actual cause of victim's death. That's not a close enough causal relationship to hold Bill Gates accountable for the actions of Bob and Mary's son, so the law requires more.

Accordingly, to prove causation in a crime, the prosecution must also prove **proximate cause** (also sometimes called "legal cause"). The proximate cause of a crime is the action most directly linked and most obviously responsible for the injury. It is a measure of probability and foreseeability. Is it probable and foreseeable that if you pull the trigger while pointing a gun at someone, they could die? Yes. Then pulling a gun's trigger while pointing the gun at someone is the proximate cause of their death.

Is it probable and foreseeable that if you introduce a couple to each other at a school dance, that someone will die? No. So it's not the proximate cause. Bill Gates is safe from a conviction for this crime.

When we say the harm must be the foreseeable consequence of the action, it doesn't mean that the defendant has to have been able to foresee the specific type of harm, just that the harm that occurred was within the natural and probable consequences of the act. So, it is no defense to say you intended to shoot someone but had no idea the victim would actually die. Death is a probable and foreseeable consequence of a shooting, even if the one pulling the trigger doesn't foresee the result.

An act can also be the cause of injury even if all it does is hasten an inevitable result. If someone is terminally ill and will die within months, killing that person is still murder.

HARM

The final element of a crime is harm, or damages. Except for inchoate crimes,[26] all crimes must cause some harm.

The degree of harm varies greatly. It could be harm to a specific victim (e.g., battery), to property (vandalism), or to society as a whole (possession of illegal contraband). In some sense, all crimes offend society because society seeks for a rule-abiding populous.

The harm element of a criminal case is much simpler than in a civil case, where the damages element is designed to calculate the monetary equivalent of the loss. In criminal law, the monetary value of the loss is often not important. Instead, the severity of the harm determines in part what the charges will be (think attempted murder v. murder or assault v. battery), and if convicted, also becomes part of the sentencing calculus.

---

[26] Inchoate crimes are discussed in further detail beginning on page 31.

CHAPTER THREE

Accomplice Liability

SO FAR, WE HAVE EXPLORED how laws are created and organized, and different theories of punishment. We have discussed the mental and physical elements of a crime and how those relate to the harm done. Now we're going to look at how the people committing the crimes are classified and how criminal liability can expand beyond just the principal actor in a crime.

PARTIES TO A CRIME

In Steven Soderbergh's 2001 movie, *Ocean's 11*, the protagonists, Danny Ocean and Rusty Ryan, put together a team to pull off a three-casino, $160,000,000 heist. Linus is a pickpocket, Frank and Saul are con men, twins Virgil and Turk are mechanics, Livingston is a surveillance expert, Basher knows explosives, and "The Amazing" Yen is an acrobat.

Together, they perform a complicated but entertaining con, successfully relieving Ocean's enemy and casino-mogul Terry Benedict of a fortune.

Suppose the eleven of them were caught and charged with breaking and entering, forgery, wire fraud, grand larceny, assault and battery, burglary, etc. To what extent is each of them responsible for each crime? How do you sort that out based on each of their individual roles? And what if some of them didn't actually do anything illegal, but their actions were nonetheless essential to the heist?

The law has changed in this respect. In most jurisdictions, what each participant does or does not do in relation to the crime used to play a much larger part in determining who was responsible than it does today.

Those primarily responsible for the crime and are present when the crime is committed are called **principals**. The law used to distinguish between two types of principals. **Principals in the first degree** were the ones who actually committed the crimes. **Principals in the second degree** were those, also present at the scene of the crime, who helped or encouraged the principals in the first degree.

Those not at the scene of the crime but who nonetheless participate in some way are called **accessories**. They are further distinguished based on whether they help before or after the crime, respectively, **accessories before the fact** (like providing access codes and security guard schedules) and **accessories after the fact** (like someone who gets rid of the murder weapon after the crime is committed).

These distinctions were important to determine who could be held accountable for what. And if the principal was not or could not be convicted, then the accessory couldn't be, either.

Most jurisdictions now have done away with these distinctions and now treat all principals and accessories before the fact as principals. This is called accomplice liability. Whether you're a principal (the one who robs the bank) or an accomplice (the getaway driver), your criminal liability could be the same.[27]

Accessories after the fact are still treated separately and are not punished for the original crime, but rather for obstruction of justice, aiding escape, or harboring a fugitive. To be liable as an accessory after the fact, the underlying crime must have been a felony and completed when the aid is rendered.

That means that all of Ocean's ten co-felons could be responsible for all the crimes that any of them committed during the heist.

Actus Reus and Mens Rea in Accomplice Liability

We learned in Chapter 2 that each crime must have an *actus reus* and *mens rea* component. With group crimes, even if only one member of the group engages in the criminal act, or *actus reus*, all other parties who have guilty intent, or *mens rea*, will be criminally liable. *Mens rea* must be analyzed separately for each participant, and each participant who has the same requisite intent required to convict the principal is guilty of the crime.

To measure the guilty mind of the apparent accomplice, we look for some affirmative assistance like an encouraging word or physical act. They could also be liable as an accomplice where there is a legal duty to act, which they fail to perform, which allows the crime to occur (like a security

---

[27] The exact scope of accomplice liability is discussed in the next section.

guard who looks the other way during a break in).

SCOPE OF LIABILITY

Determining to what extent an accomplice to a crime is liable for the crimes committed by the principal can be complicated, and there is no simple rule that fits all situations cleanly.

As discussed above, generally, a person is liable for the crimes he did himself (as the principal) and for those crimes he encouraged or for which he provided aid (as an accomplice). But he is also criminally liable for the other crimes committed by the principal that were not encouraged or contemplated, as long as they were probable or foreseeable. This is an example of the scope of liability being broadened.

In the movie, *Baby Driver*,[28] Baby is a kid who incurs a debt he can't repay when he tries to steal a car owned by Doc, a heavy-handed kingpin and conman. He gets involved as a getaway driver for Doc's various bank-robbing crews to pay back his debt. During a particular heist, one of the crew, Bats, kills a security guard. Baby is disgusted, and when his co-felons get in the vehicle, he refuses to drive.

Is Baby liable for the murder, or just the robbery? It comes down to whether the murder was a "natural and probable consequence" of the commission of the underlying crime. And during an *armed* robbery, murder probably is. Which means he could be convicted for both armed robbery and murder. The law is designed to strongly discourage people from participating in felonies in even minor ways.

In the 1990s, a 19-year-old boy with no criminal history acted as an interpreter for his Spanish-speaking buddy.

---

[28] ©2017.

Ordinarily, there would be nothing criminal about converting words from one language to another, except that this young man was translating the words necessary to effectuate a drug deal between a seller and buyer who didn't speak the same language. For his complicity in the crime, the 19-year-old was sentenced to ten years in federal prison with no chance of parole.

One New York Times Article laid out a number of examples of accomplice liability:

> *a person can be convicted of possession with intent to distribute even if he or she never touched the drugs; a person can be convicted of drug trafficking for merely referring customers to a seller; giving a known drug dealer a ride or giving a party where others sell drugs can result in prosecution, and such offenses can be felonies and result in long prison sentences."*[29]

There are also situations where the scope of liability is narrowed rather than expanded where a criminal statute has as its purpose the protection of the person who would otherwise be considered an accomplice. If an adult engages in a sexual relationship with a consenting minor below the age of consent, that act constitutes statutory rape. The consenting minor fits the definition of an accomplice, but because the statute is designed to protect minors, even consenting ones, the minor is shielded from accomplice liability.

There are also provisions for repentant accomplices whereby they can rid themselves of accomplice liability through **withdrawal**. To be effective, the withdrawal must occur before the crime has been completed and before it reaches the point of no return for completion.

---

[29] "Lawyer Tells Youths of Drug Deal Risks," *New York Times*, August 23, 1998.

What constitutes a successful withdrawal depends on how much participation and encouragement has been rendered. If the accomplice has merely encouraged the crime, she can withdraw by repudiating the encouragement (or discouraging the crime).

If she has provided aid or help, she must neutralize the aid she's provided. This may require calling the police.

CHAPTER FOUR

Inchoate Offenses

STEVEN SPIELBERG'S *MINORITY REPORT*,[30] tells the story of three psychics, called "precogs," who assist Washington D.C.'s pre-crime unit to stop murders before they happen, which reduces the capital's capital crime rate to zero. (See what I did there?)

The movie gets interesting and poses some thought-provoking philosophical questions when the precogs previsualize the murder of a Leo Crow at the hands of the Chief of the Pre-Crime Unit, John Anderton—thirty-six hours before it happens. Anderton, who does not know Leo Crow, insists that he would not murder him. When his pleas are disregarded, he flees the department as a fugitive and sets out to prove that sometimes the precogs are wrong.

One of the major themes of the film is that someone can

---

[30] ©2002. It was based on a 1956 short story of the same name by Philip K. Dick.

be convicted of a crime they haven't yet committed. And although the film is science fiction, in reality, the law does allow for criminal punishment of certain incomplete crimes. These are called **inchoate**, or incomplete, offenses. Unlike *Minority Report*, however, to be liable for an inchoate crime, there must have been some action taken towards the completion of the crime.

If you think back to the purposes of criminal law, you may realize that many of those purposes can be accomplished and are fitting even for crimes that are not complete, including incapacitation, deterrence, rehabilitation, and education.

The three inchoate crimes are attempt, solicitation, and conspiracy.

ATTEMPT

They say there's no harm in trying, but that's not exactly true if what you're trying is illegal, because when it comes to crimes, you can be punished even if you fail. The first inchoate crime, attempt, is fairly self-explanatory.

An **attempt** generally occurs when there is both (a) intent to commit a crime and (b) an overt act towards the commission of the crime. However, (c) for reasons outside the criminal's control, the crime is not completed.

Note the parallels between the elements of attempt and the elements of every other crime: there is a *mens rea* requirement (intent/recklessness) and an *actus reus* requirement (overt act beyond mere preparation).

Mens Rea in Attempt Cases

In most attempt cases, specific intent is required, so the person attempting a crime must have actually intended for the crime to occur. Thus, in most jurisdictions, to convict someone for attempted murder, you must prove that they actually intended the victim to die.

In other cases, intent is not necessary, and recklessness is enough. In an attempted rape, where the would-be rapist's intent is sexual intercourse, it is not a defense for him to say he wasn't sure whether he had his victim's consent. He does not have to have intended sex without consent; it is enough that he intended sex and was reckless about her consent.

Actus Reus in Attempt Cases

The hardest question to answer when it comes to the overt act in an attempted crime is whether the act was merely preparatory (and not enough to qualify for attempt) or whether it went beyond preparation and made up a substantial step towards the completion of a crime.

It will often come down to a question of proximity—how close was the overt act to the completion of the crime? Someone who is apprehended after acquiring floor plans of a bank and casing it to identify camera dead zones and security guard routes is going to be harder to convict for attempted robbery than someone who has walked into a bank wearing a mask.

Because preparation is arguably on a continuum, different jurisdictions have developed different legal tests designed to determine what actions are enough to constitute attempt. The details of these various tests are beyond the scope of this text, but in some jurisdictions, proximity is the principal test, but in the majority of states, they will look to see if some substantial step has been taken towards the commission of the crime.

35

Impossibility Defense

It is not a defense to say that the attempted crime could never have been accomplished. This is known as a **factual impossibility**. If I try to light your swimming pool on fire, I can still be charged with attempted arson even though your swimming pool has been fireproofed by virtue of it being a pool.

Legal impossibility, by contrast, is a defense and occurs when the attempted act is not actually illegal, even if I had guilty intent. For example, if I attempt to light your trash on fire, and everything about the way I am lighting your trash on fire is legal, even though I thought what I was doing was illegal, then I have a complete defense to attempted arson.

Would It Kill You to Prepare a Shorter Speech? Actually, Yes.

In October 1912, John Schrank tried but failed to assassinate then-president Theodore Roosevelt. Schrank, a saloon owner, shot at the President and hit him in the chest. But because President Roosevelt had a fifty-page speech and a steel case for his glasses resting in his breast pocket, the bullet didn't make it past the President's skin. Undeterred by this attempt on his life, President Roosevelt delivered a 90-minute speech (I understand it didn't suffer for length even though the bullet made his written speech resemble swiss cheese) and found it appropriate to alter his opening remarks: "Ladies and gentlemen, I don't know whether you fully understand that I have just been shot; but it takes more than that to kill a Bull Moose." Daaaang.

This cowardly attempt to take a life (and aren't they all?) illustrates perfectly the justification for meting out punishment to people who don't actually succeed in committing the crimes they intend. Schrank couldn't go down for murder, but it was an easy case for attempted murder (it's also a battery, as you'll see in Chapter 5).

His Hopes Were Squashed

A man in 2014 armed with nothing more than his wits and his weapon of choice—a cucumber wrapped in a black sock—demanded cash from a bookstore worker in Glasgow, Scotland. An off-duty police officer who witnessed the event tackled him, and he dropped his cucumber.[31] Weaponless, he realized he was out of options and surrendered. As armed robberies go, he is innocent. The most he'll be able to brag about in prison is attempted robbery. And he'll always be wondering if things might have turned out differently if he had had a tighter grip on that cucumber.

Do You Have My New Address?

Another example of attempted robbery revealed itself in Wales that same year. After visiting the local branch of his bank to change his address, a soon-to-be-criminal noticed the amount of cash the teller had on hand and came up with a genius plan to acquire it. He left the bank, grabbed a disguise Clark Kent would approve of—sunglasses—put socks on over his shoes for good measure, and 30 minutes after his first visit, went back into the bank with a knife and demanded cash. The teller refused, and he ran away.

I imagine the teller, if questioned, would say something like this: "I could not identify the man from his feet or eyes because he had the foresight to cover them, but luckily, he had the same face, clothes, and voice that he had when he came to my window a few minutes before. His transaction was still in my recent history, so I pulled up his name and new address

---

[31] Because that cucumber had never known vinegar, any pickle jokes here would be premature.

to give the police." This attempted robbery earned the quickly apprehended Dean Smith a two-and-a-half-year sentence.

SOLICITATION

Solicitation, like attempt, is an inchoate crime. **Solicitation** occurs when one person, through the use of words, encourages someone else to commit a felony with the specific intent that the person solicited actually commit the crime. It only qualifies as solicitation if the solicited person does not commit the crime. Once the crime is committed, or agreed to be committed, or attempted, the solicitor is no longer a solicitor but becomes a co-conspirator.[32]

The *mens rea* of solicitation is the intent for someone else to commit a crime. The *actus reus* is the words spoken to get someone to commit it.

Impossibility is not a defense to solicitation—the solicitor is charged based on the facts as she believed them to be. It is no defense to say the crime would never have occurred because the person solicited was an undercover cop.

Note to Self ... Don't Send Notes to Self Through Prison Mail

Quinton Thomas, an inmate in Boyds, Maryland who was incarcerated because could not make bail pending his trial for murder and armed robbery, learned about a loophole in the prison system: although incoming mail was scrutinized before handed over to the inmates, outgoing mail was largely left unmolested.

Thomas figured it couldn't hurt to increase the odds of an acquittal, so he sent a letter to his friend asking that he kill the

---

[32] Read more about conspiracy starting on page 37.

witnesses who were to testify against him at his trial. Thomas was wrong. It's never a good idea to ask others to murder, and it's a doubly terrible idea to send a letter without making sure you have the right address and correct postage. Otherwise, it will get returned to the sender, and as incoming mail, it will be thoroughly read.

Oops.

In addition to his other crimes, Thomas was convicted of solicitation to commit murder.

Being Cheap Can Be a Crime

David Magnus approached a woman in Casper, Wyoming to offer a trade. If she would have sex with him, he would give her a quarter pounder and fries. Intrigued, the woman pulled out her handcuffs. And her badge. And arrested him for solicitation.

If Magnus had been just a little bit cheaper and just asked for sex for free, he would not have been committing any crime.

CONSPIRACY

The last inchoate crime is conspiracy. A **conspiracy** is an agreement between two or more persons to do something unlawful where at least one of the co-conspirators commits an overt act in furtherance of the conspiracy.

The *mens rea* of a conspiracy is the intent to commit a crime—the criminal purpose behind the conspiracy. Each co-conspirator must have two different intents: the intent to agree, which is inferred from the coconspirators' conduct, and the intent to commit the underlying crime (to achieve the objective of the conspiracy). Only those members of the conspiracy who have both these intents can be guilty of conspiracy, though not every member of the conspiracy must know every detail of the crimes to be committed.

The *actus reus* of conspiracy also has two components: the agreement and the overt act. The agreement may be inferred based on circumstantial evidence.

For any given co-conspirator to be liable for conspiracy, all of the following must be true:

1) That co-conspirator must intend to agree;
2) That co-conspirator must actually enter an agreement;
3) That co-conspirator must intend for the underlying crime to be committed; and
4) Any co-conspirator must commit an overt act in furtherance of the conspiracy.

Unlike attempt, in conspiracy, the overt act need not be in close proximity to the crime or a substantial step towards completion of the crime. It doesn't even have to be illegal. It merely has to further the purpose of the criminal agreement, and usually, mere preparation will be enough.

With attempt and solicitation, the commission of the crime renders moot a charge for attempt or solicitation (the doctrine of **merger**). It is possible, however, for a conspirator to be charged with both conspiracy and the underlying crime.

Additionally, similar to accomplice liability, every member of a conspiracy is guilty of the crimes of their co-conspirators, even if they did not participate in the crimes, as long as (1) the co-conspirators committed the crimes in furtherance of the objectives of the agreement, and (2) the crimes were the natural and probable consequence of the conspiracy. This is known as *Pinkerton* liability.[33]

Walter and Daniel Pinkerton, brothers, were convicted of conspiracy to violate the tax code even though Walter was the only one who had actually committed any tax crimes. There was no proof Daniel did anything other than conspire with Walter to commit the crimes. He was nonetheless convicted for violations of the tax code because he was a member of the conspiracy when Walter broke the law, and Walter's violations were a foreseeable consequence of the conspiracy and furthered its objectives.

Conspiracy Defenses

Because the crime of conspiracy occurs as soon as there are an agreement and an overt act, it is not a defense that the criminal objective would have been impossible to achieve, and withdrawal is also not a defense to conspiracy itself. Withdrawal would, however, be a defense to the underlying crimes committed by the co-conspirators, as long as the withdrawing party notifies all co-conspirators of the withdrawal with sufficient time that they can abandon their plans.

Idiots in Disguise

Iowans and co-conspirators Joey Miller and Matthew McNelly were not professional burglars. But they wanted to be. They had no training, but they wanted to at least look the part, which meant wearing masks (think Beagle Boys or the Hamburglar). With no prior scores to fund a proper burglars' toolkit, they did the next best thing and scribbled on their faces with black permanent marker.

Properly disguised, these co-conspirators went to burgle their first apartment. The front door proved to be too much

---

[33] *See Pinkerton v. U.S.*, 328 U.S. 640, 66 S.C. 1180, 90 L. Ed. 2d 1489 (1946).

for them, though, and they left, unable to enter the apartment—but not before having their descriptions reported to the police. A highly trained police force took no time at all to find the only two idiots in Carroll, Iowa with guilt literally written all over their faces.

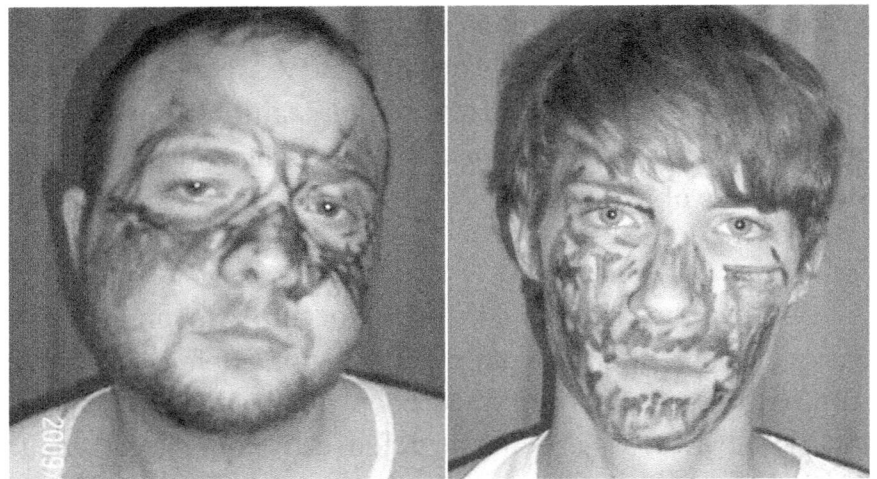

CHAPTER FIVE

Crimes Against the Person

THE EARLIEST OF ALL LAWS promoted safety and security by prohibiting interference with another's person. When Thomas Jefferson declared that we all have inalienable rights, namely, life, liberty, and the pursuit of happiness, he might as well have been describing those rights that, if interfered with, constitute crimes against the person.

Interfering with the right to life—to continue living—is, of course, some form of homicide. Kidnapping and false imprisonment interfere with one's liberty, and victims of assault, battery, and mayhem have been hampered in their right to pursue their own happiness without the threat of offensive or harmful physical contact.

One of the best scenes to come out of a James Bond movie, and perfect for illustrating a number of crimes against the person—is the scene from *Casino Royale*[34] where an unconscious Bond is carried into the dank underbelly of a ship in

---

[34] ©2006.

the Adriatic Sea off the coast of Montenegro. He is awakened, stripped naked, and tied to a chair that has had its caned seat cut out.

The sadistic Le Chiffre takes a thick knotted rope and swings it enthusiastically towards Bond's most sensitive and vulnerable and gender-defining parts. Le Chiffre's purpose is to extract an account password for his ruthless employers.

Le Chiffre then starts his bad-guy monologue. He confesses that he never understood the appeal of elaborate torture when the knowledge of how to cause a man unbearable pain is simple and was never a secret. He continues, "Of course, it is not only the immediate agony, but the knowledge that if you do not yield soon enough, there will be little left to identify you as a man."

After the first two very obviously painful blows, Le Chiffre leans in close to Bond to elicit the desired account password.

Bond, in obvious pain, manages to say, "I've got a little itch. Down there. Would you mind?"

Le Chiffre, who lacks a sense of humor, angrily gets to his feet and swings his whip again.

"No. No! NO!" Bond's cries crescendo and give Le Chiffre some satisfaction until Bond continues, "To the right! To the right!"

Le Chiffre, frustrated that Bond is not responding as he expects, once again obliges.

"Yes!" Bond screams, and Bond's delight breaks through his pain with one of the funniest lines in the movie, which is very appropriate for the scene, but just below the bounds of propriety for a textbook. Google "Casino Royale torture scene" and watch the video to get the full effect.

This little scene will serve as a backdrop to our

discussions of assault, battery, mayhem, false imprisonment, kidnapping, and homicide. Yes, homicide, because Bond's words (the ones edited out of the text) turn out to be prescient—not a minute after Bond's exclamation, Le Chiffre's boss enters the room and shoots Le Chiffre dead.

Like other crimes, each of these crimes against the person has its own set of elements, including both *mens rea* and *actus reus* components. They also vary widely from state to state and jurisdiction to jurisdiction. We attempt here to reveal the threads common among these jurisdictions.

ASSAULT AND BATTERY

Assault and battery are two terms whose legal meanings differ from the common understanding. **Battery** is an unlawful harmful or offensive application of force. Slapping. Pushing. Shooting. Unwanted sexual contact. These are all forms of battery.

Assault is either an attempted battery or the intentional creation of apprehension of a battery. The distinguishing feature of battery is physical contact: assault is independent of any physical contact.

Battery

In most jurisdictions, a battery does not require actual injury if there is some harmful or offensive physical contact.

The *mens rea* of battery also differs from jurisdiction to jurisdiction. It's always enough if the batterer intends to injure or touch offensively. In some jurisdictions, though, even without intent, it's still a battery if the harmful or offensive touching occurs as a result of criminal negligence.

And intent can be transferred. The doctrine of

transferred intent says that if I intend a battery against one person, but inadvertently or accidentally commit a battery against another (e.g., I throw a rock at person A, but accidentally hit person B), I've still committed a battery. This is true even though I did not have the intent to make physical contact with the victim.

The *actus reus* of battery is the application of force—the physical contact or touching. Skin-to-skin touching is not required. It's enough that contact is made with an object closely associated with the body.

Contact also does not have to be immediate. If someone puts a series of actions in motion that results in physical contact, like introducing poison to someone's food, shooting an arrow from a bow, or siccing a junkyard dog, that is still an application of force.

The contact does not have to do actual harm as long as it is offensive, so unwanted fondling or blowing smoke in someone's face would qualify.

Where the batterer uses a deadly weapon, inflicts serious bodily injury, or the victim is a police officer, woman, or child, some jurisdictions will bring charges of **aggravated battery**.

In our James Bond example, Le Chiffre committed battery every time he whacked Bond with the knotted rope—he intended to make (*mens rea*), and actually did make (*actus reus*), harmful (and offensive) physical contact with Bond.

Assault

Assault is one of those words that is used wrong all the time. In Adam Sandler's *Billy Madison*,[35] the titular character, a 27-year-old screw-up, goes back to school to prove to his millionaire father that he's qualified to take over the family

---

[35] ©1995.

business. While on a bus during a third-grade field trip, one of the students dares Billy to touch their teacher inappropriately. Billy famously responds, "That's assault, brotha!"

Actually, in most states, it's not.[36] An assault ends where a battery begins—if there is actual touching, it's no longer assault. Like a caterpillar emerging from a chrysalis as a butterfly, once contact is made at the tail end of an assault, it has now metamorphosed into battery. Except a butterfly is the caterpillar's beautiful counterpart while battery is much uglier than assault.

An assault occurs in one of two non-contact situations: (1) when someone attempts to make harmful or offensive physical contact (i.e. attempts a battery), or (2) when someone intends and uses more than words to put their victim in reasonable apprehension of immediate bodily harm (i.e. makes them think a battery is about to occur, even if it is not).

The first kind of assault is basically a swing and a miss. It's like any other attempted crime—there is intent to commit the crime, but because of some circumstance outside the criminal's control, it does not happen.

The second kind of assault is what I call the "flinch assault." You have no intention of hurting anyone, but you want them to think you do. That's still a crime because the law says we have a right not to be fearful of imminent physical harm.

Either can occur when physical contact occurs. Two swing a bat and hit someone is assault *and* battery. It's assault up until contact is made (assuming the victim is aware of the pending contact and apprehensive of it) and then it is battery. A person can be charged with both.

Was there anything about the *Casino Royale* torture scene

---

[36] A minority of jurisdictions use the term "assault" to refer to what most others understand to be a battery.

that constituted assault? Absolutely.

When Le Chiffre swung the rope towards Bond, that was an assault only until the moment when the rope made contact. Then it was a battery. If you watch the full scene there are other parts not described above that would constitute the crime of assault only.

MAYHEM

Mayhem is a funny word. If the word means anything to you, it's probably from the invincible guy from the All State commercials who promises that if you go with their insurance, you'll be "better protected from mayhem ... like me."

But mayhem has much older roots than insurance companies pretending they care about the people they insure.[37] As a crime, it actually originated in Great Britain and was enacted to protect the king's soldiers so they could defend themselves and the crown. To that end, mayhem consisted of damage to an eye or a limb—body parts that were necessary for combat. If you lost something less integral to fighting, like your nose, it was not mayhem because you could still swing a sword or throw a rock with the same accuracy.[38]

Mayhem eventually evolved into any maiming, mutilation, or dismemberment of a body part. Some jurisdictions retain this crime, while many others have done away with it, calling this type of harm aggravated battery.

---

[37] Pardon my cynicism, but it's well earned. If you'd like to offer a defense for insurance companies, let's start a dialogue. My email is zach@paralegaltrainer.com.
[38] Mayhem has some interesting root words, including Anglo-French *mahaim*, meaning "maim," and Middle High German *meidem*, meaning a castrated male animal (like a gelding).

My favorite recent example from the media of mayhem[39] comes from the fifth episode of the first season of Seth MacFarlane's *Star Trek* parody, *The Orville*. It involves two of the bridge crew.

Isaac is a member of a race of synthetic, mechanical life-forms (read: he and his kind are robots). He joined the Orville's crew to study humans, which his race considers an inferior life form. Because he is a robot, he has no sense of humor and voices his confusion at why humans like *Seinfeld* reruns.

Gordon Malloy, on the other hand, is a red-headed member of the inferior species—humans—who finds *Seinfeld* hilarious, loves a good prank, and wants to teach Isaac about humor. He decides to play a prank and attaches Mr. Potato Head parts to Isaac's metal head, to the great delight of the bridge crew. Gordon explains to Isaac that this is what is known as a practical joke, and he encourages Isaac to come up with one on his own.

Thinking he understands human behavior, Isaac approaches Gordon in his sleep and amputates his left leg. Gordon realizes he has been maimed when, getting out of bed in the morning, he loses his balance and falls over. He hops onto the bridge understandably angry, unable to see the humor in Isaac's practical joke. Luckily, they have the technology that allows him to grow back his leg in about a day.

Amputating someone's leg, even as a joke (maybe especially as a joke), is criminal mayhem.[40]

If you want an example of attempted mayhem, go watch the *Casino Royale* torture scene. Le Chiffre is about to commit

---

[39] Yes, I rank mayhem examples in order of preference. Don't you?
[40] Unless you're a doctor and the procedure is medically necessary.

criminal mayhem when his boss interrupts him with a bullet.

HOMICIDE

Homicides have become so common they are no longer headlines, especially as we can't go too long without hearing about another mass shooting by some crazy psychopath. Assault, battery, kidnapping, false imprisonment, and mayhem, by these or other names, are also common in both real life and in drama.

Sex crimes, a special category of crimes against the person, will be treated in Chapter Six.

One ambitious journalist gathered data about all the deaths of major characters on a single season of television across all channels.[41] She uncovered 242 deaths:

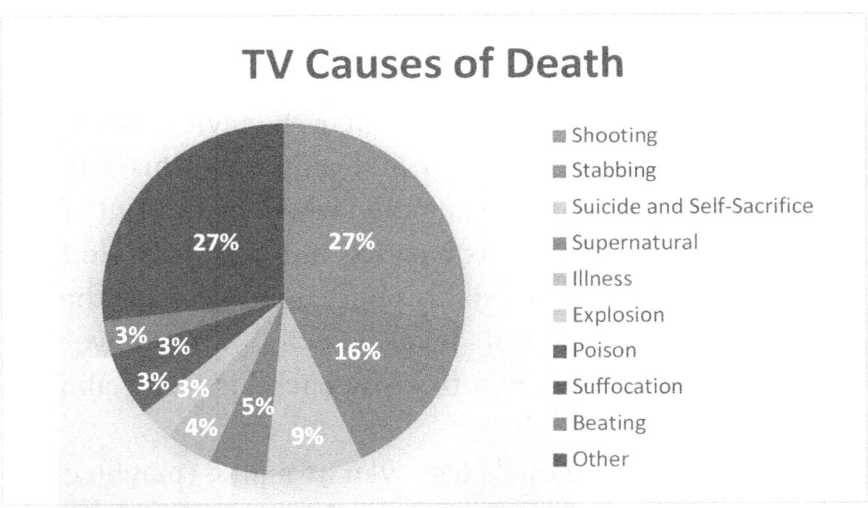

These deaths are only a tiny percentage of actual death portrayed on television and don't count the characters

---

[41] *See* Caroline Framke, "All the TV Character Deaths of 2015–'16 in One Chart," *Vox*, June 1, 2016, available at https://www.vox.com/a/tv-deaths-lgbt-diversity (last accessed April 10, 2018).

already dead when the episode begins or the so-called "red-shirts"—television characters whose role in the episode is to create a sense of danger to the main characters by dying themselves.

As a crime, homicide takes many different forms. Generally speaking, **homicide** occurs when one person kills another. Historically, homicides were divided into three categories: justifiable, excusable, and criminal.

When the killing of a human being is authorized by law, we say it is **justifiable homicide**. Soldiers killing enemy combatants during a battle and capital punishment are examples.

Excusable homicide, on the other hand, applies when the killer has a legal defense to the killing, like insanity, self-defense, or defense of others.

The most serious of homicides are the **criminal homicides**, which are, as you might have guessed, not justified, not excusable, and constitute a crime. Criminal homicide themselves comprised three subcategories: murder, voluntary manslaughter, and involuntary manslaughter.

The key ingredient in **murder** is "malice aforethought," which requires either express malice, i.e., intent to kill, or implied malice, which is inferred either from an intent to inflict serious bodily harm (e.g., you hit someone in the head intending to knock them out, but they die); reckless indifference to the high risk of fatal harm, known as the "abandoned and malignant heart" (e.g., you fire shots in the air above a crowd); or the intent to commit a felony, known as the felony-murder rule, discussed in greater detail below (put simply, you are in the middle of a felony, and you or one of your co-conspirators causes someone's death).

The default for murder is murder in the second degree, or **second-degree murder**, which occurs due to "outrageous disregard for life" or "wanton recklessness." **First-degree murder** requires the prosecution to prove that the killing was both deliberate (the decision to kill was made dispassionately) and premeditated (the killer reflected on the killing beforehand).

The **felony-murder rule** is a special category of murder where an accidental death occurs during the commission of dangerous felonies. To be guilty of murder under this rule, you must have actually committed the felony, that felony must be the cause of death, and the risk of death must be foreseeable. Different states vary the rule, and some have abolished it altogether.

When the felony-murder rule is combined with *Pinkerton* liability of co-conspirators, it's application can be rather broad—all co-conspirators can be criminally liable for a death caused by any one of the other co-conspirators. The lesson here is not to participate in dangerous felonies. Don't do it alone, and don't do it as a group. Non-dangerous felonies are also really good to avoid.

Down a rung from murder in the ladder of degrees of criminal homicide is **voluntary manslaughter**, which is intentional killing accompanied by adequate provocation. To qualify for the lesser offense of voluntary manslaughter, the person charged with murder must be able to prove that two things are true both for a reasonable person and for the defendant: (1) that the victim provoked the defendant to such a degree that the ordinary person would lose control, and the defendant actually did lose control; and (2) that there was not enough time for a reasonable person to cool off between losing self-control and doing the killing and that the defendant didn't actually cool off.

The prototypical example of voluntary manslaughter is a

husband or wife who walks in on their cheating spouse *in flagrante delicto*, and immediately kills one or both of the cheating pair.

In the movie, *A Christmas Story*,[42] a coonskin-cap-wearing bully named Scut Farkus constantly antagonizes the film's protagonist, Ralphie Parker. Day after day, Scut calls him names, chases him, and makes him say "uncle." Ralphie finally snaps and tackles Scut and begins relentlessly beating Scut in a blind fury. The beating goes on far longer than it needs to, and Ralphie's mom eventually arrives and stops the beating. Scut himself survives, but if his pride were a person, this would be a perfect example of voluntary manslaughter. Because there is no way his pride survived being beaten that badly by a nerdy boy.

Further down the homicide ladder is **involuntary manslaughter**, which is a death resulting from criminal negligence or during the commission of a misdemeanor. In criminal negligence cases, the deviation from the behavior of a reasonable person must be greater than in civil negligence cases. And if the death occurs during a misdemeanor, it usually requires that the misdemeanor be inherently wrong (*malum in se*), or if not, the death must be foreseeable.

In some jurisdictions, a death that results from driving a vehicle recklessly or while intoxicated will result in a charge for vehicular or intoxication manslaughter. That was the case for Ethan Crouch, who on June 15, 2013, as a sixteen-year-old rich North Texas kid, stole two cases of beer from Walmart, loaded up his dad's pickup truck with seven teenage passengers, drank to the point where his blood-alcohol was three times the legal limit (0.24%), then swerved off a two-lane

---

[42] ©1983.

country road at 70 miles per hour (in a 40 mph zone), careening into an SUV, which in turn hit a parked car, which in turn hit another oncoming car.

All told, four people died, and one of Ethan's passengers was paralyzed.

Charged with four counts of intoxication manslaughter, Ethan's defense, buffeted by an expert psychologist, was that his wealth made him irresponsible, a term he called "affluenza." He was ultimately convicted, and the judge sentenced him to ten years' probation and rehabilitation therapy.

The nation was outraged at the light sentencing, and a Tweet with a video showing Ethan in violation of his probation started a chain of events resulting in a warrant for his arrest, which caused him to flee.

Now a fugitive, the FBI got involved and found him in Puerto Vallarta, Mexico, where his mother had taken him to elude capture.[43] With the aid of cooperative Mexican law enforcement, they were both arrested and sent back to the United States. He was sentenced to two years in jail for his parole violation and was released on April 2, 2018.

FALSE IMPRISONMENT

In Dykes' Bunker

On January 29, 2013, Alabama resident and school bus driver Charles Poland, Jr. made a stop in Midland City, Alabama at about 3:30 p.m. He had made that stop countless times before in his years as a bus driver. This time would be different.

Sixty-five-year-old Jimmy Lee Dykes boarded the bus

---

[43] He has a terrible mother.

armed with a Ruger handgun and told Poland that he was going to take two boys with him. Poland stood up and blocked the aisle to protect the kids on the bus, but Dykes shot him dead and took a five-year-old boy named Ethan—a random victim. The captor took his captive to an underground bunker at his property.

From there, Dykes called 9-1-1 to open negotiations with the FBI. Negotiations continued for six days, with Ethan kept in the bunker the entire time. On day six, the FBI sensed that time was running out and breached the roof of the bunker then tossed in some stun grenades. In the end, Dykes tried to shoot the federal agents but was killed. Ethan was rescued and returned to his family.

The Legend of DB Cooper

On November 24, 1971, a man whose true name was never discovered purchased an airline ticket at the Portland International Airport. He paid $20 cash for the thirty-minute flight to Seattle and gave a false name, Dan Cooper.

Once in the air, Cooper notified the stewardess that he had a bomb, and he relayed his demands, which the pilot, in turn, relayed to the authorities: He wanted $200,000 in "negotiable American currency," four parachutes, and a fuel truck ready to refuel the plane in Seattle.

While the Seattle Police and FBI gathered the chutes and cash, the plane circled Seattle for two hours and then landed once the authorities told Cooper that his demands had been met. They had acquired the chutes and had $200,000 in $100 bills, all of which had been photographed. The passengers were given a false, unalarming reason for the delay. The cash and parachutes were delivered, after which Cooper told all but the cockpit crew to exit the plane.

Cooper gave the crew very specific instructions about their travel: they were to fly towards Mexico City (with a refueling stop in Reno) at the slowest speed possible without stalling the aircraft, at a maximum altitude of 10,000 feet, the wing flaps lowered by 15 degrees, the landing gear deployed, and the cabin unpressurized.

Once they had taken off again, three military aircraft were deployed and followed the plane where they could not be seen. Cooper ordered the flight crew to get in the cockpit and remain there with the door closed. When the plane landed in Reno two hours and twenty-five minutes after it left Seattle, local and federal authorities surrounded the plane, but Cooper was no longer there. The jets following him never saw him jump (though to be fair, it was nighttime and cloudy).

No one ever found him, his body, or discovered his true identity. It is unknown if he even survived. Besides $4,000 in ransom money buried in a shallow riverbank nine miles downriver from Vancouver, Washington, found by an eight-year-old boy in 1980, none of the rest of the ransom money was ever found. The FBI kept an active investigation open until 2016.

The media mistakenly called the man D.B. Cooper, which is how he has been known ever since.

What these two stories have in common is that in both, the perpetrator engaged in (1) unlawful (2) confinement of a person (3) without consent, or **false imprisonment**.

It does not have to be a hostage situation for these elements to be met. False imprisonment also includes forcing a victim to remain when he wishes to leave or to go somewhere he doesn't want to go. When Biff Tannen's cronies threw Marty McFly into the trunk of a Cadillac, that was false imprisonment.[44] When the Nazis used some thick rope to tie Henry Jones Sr. and Henry "Indiana" Jones Jr. to a pair of chairs, that was false imprisonment.[45] Han Solo in carbonite?[46] Princess Leia in a collar and chains keeping her in close proximity of Jabba the Hutt?[47] Definitely.

KIDNAPPING

At one point in history, in 1927, Charles Lindbergh was the biggest celebrity in the world when he won the $25,000 bounty for successfully completing the first solo transatlantic flight—a 33.5-hour, 3,600-mile flight from New York to Paris.

Five years later, Lindbergh would experience every parents' worst nightmare when his 20-month-old son, Charles Jr. was kidnapped from the crib in his second-story bedroom.

Lindbergh found a poorly written ransom note taped to the windowsill in his son's room and pieces of a wooden ladder under his son's bedroom window. The note demanded $50,000 and told Lindbergh to await further instructions.

There were several communications exchanged, but ultimately, Lindbergh delivered the ransom in a wooden box 32 days after the boy was taken. The $50,000 was comprised of recognizable gold certificates and bills whose serial numbers had been recorded.

Charles Jr. was never returned to his family. Instead, his lifeless body was found 73 days after the kidnapping.

Kidnapping statutes vary widely from jurisdiction to jurisdiction but generally include unconsented seizure,

---

[44] *Back to the Future*, ©1985.
[45] *Indiana Jones and the Last Crusade*, ©1989.
[46] *The Empire Strikes Back*, ©1980.
[47] *The Return of the Jedi*, ©1983.

confinement, or movement of one person by another by use of force, threat, or deception. Some jurisdictions recognize the crime of aggravated kidnapping where certain aggravating circumstances are present, like if there is a ransom demand, the kidnapping occurs so another crime can be committed, it is done with the intent of inflicting harm or a sexual crime, or the victim is a child.

In the Lindbergh case, the police created 250,000 flyers with the serial numbers from the ransom bills and distributed them to local businesses, but that wasn't how the kidnapper was caught. Franklin Roosevelt discontinued gold certificates and ordered that they all be exchanged for other bills by May 1, 1933. The compliant kidnapper exchanged his gold certificates, which were identified as part of the ransom payment. Kidnapper Bruno Richard Hauptmann was identified by the bank, arrested by police, charged and tried by the prosecutor, convicted by a jury, sentenced by a judge, and executed by electrocution. It was the kidnapping that called the world's attention to the crime, but it was the capital murder conviction that would ultimately claim Hauptmann's life.

CHAPTER SIX

Sex Offenses

ALTHOUGH SOCIETY'S VIEWS OF SEX have consistently changed over the course of human history, included in even the most ancient of laws—written down over 4,000 years ago—were prohibitions against unconsented sexual contact (though they were nowhere as expansive or uniformly applied as they are now).

RAPE

Recall that battery is the unconsented unlawful harmful or offensive application of force. Rape in all its forms fits this definition and would also qualify for aggravated battery. However, because as an act, rape is universally condemnable and in most cases much more damaging to the victim, it is its own crime, separate from battery.

In the United States, we condemn forcible sex in all cases, though that wasn't always true. Before 1975, if a man forced his wife to engage in intercourse, it was not considered a crime in any state. North Dakota became the first state to make it a crime for a man to rape his wife. Since 1975, every state has made marital rape illegal, ending with North Carolina in 1993.

Until 2012, the federal definition of **rape** was the same as it had been since 1927: "the carnal knowledge of a female, forcibly and against her will." Thus, only women could be victims of rape, and it was limited to penetration of the female sex organ by the male sex organ. Now the definition is much broader and is no longer gender specific, nor limited to vaginal penetration: "The penetration, no matter how slight, of the vagina or anus with any body part or object, or oral penetration by a sex organ of another person, without the consent of the victim."[48]

Consent in Rape Cases

Effective consent for sexual contact is a defense to the crime of rape. If the rape is forcible, there is never effective consent. The same is true if consent is given as a result of a threat of serious and immediate bodily harm.

To what extent consent obtained by fraud is effective consent has been of great concern to courts and legislatures, and the laws in that respect vary among jurisdictions and have evolved over time.

California's laws on consent-obtained-by-fraud are instructive. In California, like much of the rest of the country,

---

[48] The United States Department of Justice Archives, January 6, 2012, "An Updated Definition of Rape," available at https://www.justice.gov/archives/opa/blog/updated-definition-rape (last accessed April 13, 2018).

fraud came in two flavors, one of which vitiated consent, and one of which did not: "**fraud in fact** occurs when the defendant obtains the victim's consent to an act but then engages in a *different* act. **Fraud in the inducement** is committed when the defendant uses misrepresentations to gain the victim's consent to an act, and then performs that *same* act."[49]

Consent obtained by fraud in fact is not consent, and the resulting intercourse is rape. Such was the case when a doctor treated women for menstrual cramps by having them change into a dressing gown, turn away from the doctor, and bend over a table with their feet apart. The doctor would insert a metal instrument as part of the "treatment," then remove it and insert his penis. The women consented to the "treatment," but were not aware it involved intercourse. The doctor was convicted of rape because the consent was ineffective—the women were consenting to a different act than what the doctor performed.[50]

Courts have historically been split about whether consent to have sex obtained by fraud in the inducement is rape. This was true in a case where a man, impersonating a woman's husband, had intercourse with her, and she consented, but only because she believed him to be her husband. Some courts have been reluctant to call this rape, though it could qualify as some other crime.

In a horrible case, also coming out of California, a man claiming to be a doctor convinced a woman over the phone that she had a terminal and loathsome disease she contracted from contact with a public toilet.

According to the man, there were two cures: (1) a painful

---

[49] *People v. Icke*, 9 Cal. App. 5th 138, 144, 214 Cal. Rptr. 3d 755, 761 (Ct. App. 2017), review denied (June 14, 2017) (emboldened text added; italics in original).
[50] *People v. Minkowski*, 204 Cal. App. 2d 832, 23 Cal. Rptr. 92 (Ct. App. 1962).

and intrusive surgery that would cost $9,000 that insurance would not cover, or (2) she could "have sexual intercourse with an anonymous donor who had been injected with a serum which would cure the disease."

This less painful, nonsurgical procedure would only cost $4,500, though she would only have to put $1,000 down. Believing her only alternative to either procedure was death, she rushed to the bank, withdrew $1,000, and had intercourse with the stranger to save her life.[51] The California appeals court expressed some sympathy for her, but in a split decision, decided that this did not fall strictly within California's definition of rape.

The California legislature amended the rape statutes in 2002 "to expand the circumstances under which a defendant may be prosecuted for fraudulently inducing a victim to consent to sexual conduct," and under the new statute, sex induced by fraud is sex performed without consent, i.e. rape.

There are also situations where consent cannot be given, like where the victim is unconscious, under the influence of drugs or intoxicating substances, or mentally incapacitated or disabled.

Such was the case in the famous Stanford Rape case. What made that case unique was not the nature of the crime but the leniency of the punishment.

Nineteen-year-old Brock Turner was a freshman at Stanford when in the wee hours of January 18, 2015, two graduate students found him having intercourse with an unconscious woman behind a dumpster.[52] Although Turner tried to flee, the two graduate students apprehended him and turned him

---

[51] *Boro v. Superior Court*, 163 Cal. App. 3d 1224, 210 Cal. Rptr. 122 (Ct. App. 1985).
[52] Which was probably the court's first clue that this sexual encounter was not the product of a romantic evening out.

into the authorities.

He was arrested, charged, tried, and convicted. At the sentencing hearing, prosecutors recommended a six-year prison sentence. Judge Aaron Persky sentenced Turner to six months in jail and three years' probation. Turner was released after three months, which, like the Ethan Crouch case, caused a public outrage at the light-handed sentence. According to reports, Judge Persky was generally well respected as an arbiter before he handed down this sentence. However, in what is perhaps a reflection of society's disdain for sexual crimes (and perhaps some foreshadowing of the #metoo movement), as of January 24, 2018, 94,539 voters signed a petition to recall the judge, and on the June 5, 2018, ballot, voters forced Judge Persky to turn in his robe and gavel in a recall vote that won by a 60/40 margin.

STATUTORY RAPE

In some cases, society's goals are such that we do away with the *mens rea* requirement altogether and punish someone for committing a forbidden act even if they do not have a guilty mind—sometimes the criminal may not even know they have done anything wrong.

Statutory rape is one such crime. **Statutory rape** is the consensual sexual intercourse with a person under the age of consent or a person who is mentally or physically incapacitated.

The age of consent varies state to state, and the difference in age between the two consenting sexual partners matters in most states—meaning that it is only illegal to have sex with someone younger than the age of consent if the other partner is older by a certain amount or at least a certain age.

In 33 states and the District of Columbia, the age of consent is 16.[53] In six states, it's 17[54]; and in the last eleven states, it is 18.[55]

In a minority of states, if there is a reasonable mistake about the younger person' age—like if the younger partner[56] lies about how old he or she is—that can be a defense to charges of statutory rape.

Leah Shipman, a 42-year-old high school teacher in North Carolina, was accused of sleeping with her 15-year-old student. She had a creative solution to her problem that not only pretty much confirmed that she had actually engaged in the illicit relationship, but also protected her from being prosecuted for it.

In North Carolina it is a crime (1) to have sex with someone younger than 16 (unless the age difference is 4 years or less), and (2) for a teacher of any grade K–12 to engage in sexual activity with a student, regardless of age. However, neither of these is a crime if the couple is married.

Shipman was not charged with statutory rape until the student was 17, so he was old enough to consent to sex, just not with his teacher. Shipman divorced her husband of 19 years and married the youth. Now that they were married, they could legally carry on with the sexual relationship.

Their marriage did not prevent the state from

---

[53] Alabama, Alaska, Arkansas, Connecticut, Georgia, Hawaii, Indiana, Iowa, Kansas, Kentucky, Maine, Maryland, Massachusetts, Michigan, Minnesota, Mississippi, Montana, Nebraska, Nevada, New Hampshire, New Jersey, New Mexico, North Carolina, Ohio, Oklahoma, Pennsylvania, Rhode Island, South Carolina, South Dakota, Vermont, Washington, West Virginia, and Wyoming.

[54] Colorado, Illinois, Louisiana, Missouri, New York, and Texas.

[55] Arizona, California, Delaware, Florida, Idaho, North Dakota, Oregon, Tennessee, Utah, Virginia, and Wisconsin.

[56] In some states, it is statutory rape only if the victim is female.

prosecuting her for the statutory rape for the sex they'd had when he was 15, though—at least not directly. The prosecutors needed to be able to prove the crime occurred, and there were only two witnesses to it: the two who participated in it. And now that they were married, Shipman could prevent the state from using her student/husband's testimony against her (spousal privilege), which insulated her from prosecution for the earlier crime.

It's not the most romantic way to start a marriage, but it's pretty effective as a defense. It did not keep her from getting fired as a teacher, though. Someone should have told her that the best way to avoid prosecution and keep your job is still the traditional approach: don't sleep with students.

CHAPTER SEVEN

Property Offenses

WE HOLD OUR PERSONAL RIGHTS—including our life, our health, our privacy, and our dignity—in the utmost regard, and offenses against these bodily rights deeply offend society's sense of justice. Offenses against property, though perhaps less degrading or impactful, are no less important for maintaining a sense of safety and order.

In the days of the great western expansion, Abraham Lincoln's Homestead Act provided that any man could claim 160 acres, and after five years of living on and tending the land, it would be deeded to him. Tens of thousands of families rushed westward to claim land—a place they could build a home and keep their families and possessions safe.

These settlers, not far descended from those seeking opportunity by leaving Europe for greener pastures, embodied what would become known as the American Dream—the idea that anyone, no matter their means or their background, through hard work, persistence, and creativity, could prosper and find material success.

And as property has always been associated with livelihood and earning potential, so, too, have there always been laws to protect the product of a person's labor—their property. The laws that protect a person's property are the subject of this chapter.

LARCENY

Larceny is among the oldest of property crimes. The divine command, "Thou shalt not steal," was a proscription against wrongs that include larceny. **Larceny** is a particular kind of stealing and occurs whenever one person wrongfully takes and carries away the property of another with the intent to deprive the owner of the property.

In simple terms, the *actus reus* of larceny is the taking and carrying away (asportation) of the property. The *mens rea* is the intent to steal.

Larceny is divided by degrees, based on the value of the property. **Petit larceny**, a misdemeanor, is the lesser form of larceny, which applies when the legal threshold for value has not been met. If the stolen property has a value equal to or greater than the legal threshold, it is **grand larceny**, a felony. Thus, all larceny is either petit larceny or grand larceny.

In Nevada, for example, the line between grand and petit larceny is $650 (increased from $250 in 2011).

Usually, the wrongful taking is trespassory, meaning without the permission of the property's owner. But there is a special category of larceny called **larceny by trick**, which occurs when the property owner consents to possession of the property, which consent is induced by fraud, and therefore not proper consent and still wrongful.

Build-Your-Own-Larceny-Charge

Three hungry dupes in Livingston County, New York with an idiotic criminal plan set their sights on a Build-a-Burger restaurant, but instead of building a burger, they built themselves a rap sheet that included grand larceny, burglary, and criminal mischief.

Their goal was to steal the restaurant's cash register and surveillance system. And if they had followed rule 83 of the criminal code, "Never thieve hungry," they might have gotten away with it. I've never tried Build-a-Burger's macaroni salad, but it must be something special because these thieves couldn't resist taking a huge helping in the midst of their crime.

Having gotten what they came for, the thieves fled the scene on foot, taking turns holding the cash register, surveillance equipment, and wolfing down that delicious macaroni salad—all while leaving a trail of pasta-and-mayo goodness that led the police straight to them.

Most of the property was recovered, but Build-a-Burger said the Hansel-and-Gretel robbers could keep the macaroni.

EMBEZZLEMENT

Embezzlement, which, like larceny can also be (but is not necessarily) a form of stealing, differs from larceny in a couple of respects. The biggest difference between embezzlement and larceny is that with larceny, the thief is stealing something that he or she never had any business possessing—the larceny occurs the moment they unlawfully take possession of the property. Embezzlement, on the other hand, deals with property entrusted with the embezzler—the embezzler has lawful possession of the property and then uses it in a way contrary to the parameters of the entrustment.

Another difference between embezzlement and larceny is that with larceny, the thief actually steals the property with the intent to permanently deprive it of the true owner. With embezzlement, the embezzler merely has to treat the property in a manner that violates the trust that was placed in him—he doesn't actually have to steal the money. He could give it away or destroy it,[57] and it will still be embezzlement.

Embezzlement, then, is the fraudulent conversion of someone else's property by a person in lawful possession of the property.

The *actus reus* of embezzlement is the conversion of the property.

The *mens rea* of embezzlement, interestingly, is not the intent to steal. It is the intent to defraud.

The Only Good Part of Superman III

Superman and *Superman II* were great movies. Still watchable, too, if you don't mind the literal tights.

Superman III, on the other hand, was terrible. Its creators wrote a rotten, un-Superman-like script, and tried to make up for it by casting Richard Pryor, which, as it turns out, gave us the only *Superman III* legacy worth remembering: the classic salami-skimming embezzlement scheme.

Richard Pryor's character, Gus Gorman, realizes that every time he gets a paycheck, his employer is paying him a fraction of a cent extra due to rounding, so he writes a

---

[57] Conversion involves the dispossession or destruction of property.

sophisticated computer program that takes the rounded cents from every employee in the company and pays it to him. He figures it will be a quick way to get a bit of a raise.

He then gets a bonus check for $85,789.90, which makes sense only if his employer employs hundreds of thousands of people. Also, apparently all the other employees received paychecks that included fractions of cents, or his scheme wouldn't work.

I don't recall exactly what Gus' position was with his company, but if he was handling the money—like if he was the accountant or was in charge of payroll—and had a right to it before it was distributed to the employees, then this would be embezzlement.

If, on the other hand, he was in IT or software development, which seems likely given his programming skills, and not tasked with handling everyone's money (and therefore never in lawful possession of it), then his theft would be larceny.

Either way, what he did was illegal, and he should have known better than to steal from his employer and then drive to work the next day in a shiny new Ferrari. At least that much *Superman III* got right—criminals are dumb.

FALSE PRETENSES

The third in our trilogy of theft crimes that started with larceny and embezzlement is false pretenses. Like these other two crimes, false pretenses is a form of stealing. Unlike larceny and embezzlement, however, false pretenses ends with the title of the stolen property being transferred to the thief. Put simply, **false pretenses** occurs when someone makes a false statement of fact to defraud another person, which convinces the other person to transfer title of their property to the first. In some jurisdictions, this is simply called **fraud**.

False pretenses is similar to larceny by trick. Larceny by trick is theft facilitated by fraud that results in the thief having unlawful possession of the stolen property. False pretenses, conversely, results not in mere *custody*, but actual *ownership* of the stolen property.

Inventors Beware

In 1978, An enterprising inventor in Utah invented what he believed would revolutionize the consumer music experience—an automatic cassette tape changer.[58]

At the same time, Global Marketing Services touted itself as a company specializing in the development and marketing of inventions. The inventor contacted Global, and Global promised him that for $2,000, it would do market research, create drawings, build a prototype, market it in catalogs, and use its best efforts to mass produce and promote the product. Global enhanced its reputation in the inventor's eyes by claiming that it had success marketing other big products, "like a speedometer, a fuel-saving device, and an electric car."

Convinced that Global was the right company for the job, the inventor gave them $2,000 and the designs for his invention.

Turns out Global was not in the invention business at all. It hadn't been working on a speedometer.[59] It didn't do any promoting. It merely made false promises in exchange for

---

[58] I don't expect you to know what a cassette tape is—the term was removed from the dictionary in 2011, after all. It was a plastic device that housed spools of long brown tape, which could be read by antiquated machines called "cassette players," which could convert the microscopic patterns on the tape to room-sized music.
[59] The speedometer was invented in 1888.

money.

This is false pretenses: Global made a false representation to induce the inventor to willingly give it his money, which he did.[60]

It's too bad, too. Who knows, if the automatic cassette tape changer had ever taken off, maybe we never would have needed to go to the cloud with our music.

ROBBERY

The fourth in our trilogy[61] of theft crimes is robbery. Robbery is distinguished from the other theft crimes mainly in two respects: the stolen property must be taken directly from the person (or from their presence), and the theft must be accomplished with threat or force. In fact, it is only those two elements that distinguish robbery from larceny. If the law were math, then the following equation would be true: Robbery = Larceny + person or presence + force or threat.

Robbery, then, is the taking of personal property from another's person with force or threat of harm, with the intent to permanently deprive them of the property.

The force required must be sufficient to overcome the resistance of the victim. If it a threat, it must be a grave threat, like the threat of serious bodily harm or death.

That's Not a Knife ...

The unlucky mugger in *Crocodile Dundee* brandished a wee little switchblade when he demanded that Mick Dundee

---

[60] These facts are laid out in the published case, *State v. Jones*, 657 P.2d 1263, 1264–65 (Utah 1982).

[61] A trilogy typically has only three. Google "Hitchhiker's trilogy" for at least one notable exception.

and his lady friend Sue turn over their wallets. Sue, frightened, told Mick Dundee to comply and turn over his wallet. Mick asked, confused, "What for?"

As though pointing out the obvious, Sue incredulously replied, "He's got a knife."

Mick laughs and famously responds, "That's not a knife. This is a knife." And he pulls out a large Bowie knife, which he uses to scare the would-be mugger away.

The robber intended to deprive Mick of his wallet, and he used the threat of grave force to do so. But since he was unsuccessful, he would not be charged with robbery, but attempted robbery, if anything. Mick wasn't too concerned that he got away, so he probably didn't care to report him. "Just kids having fun," he said with a genuine smile as the mugger turned tail and ran.

EXTORTION

With, embezzlement, larceny by trick, and false pretenses, the victim gives up the property willingly, but only because they don't have all the facts. If they knew the true motives of the person to whom they are handing over their property, they would never hand it over. Extortion also involves the voluntary relinquishment of property, usually money, but the victim knows they're being taken, doesn't want to pay, but does it anyway. **Extortion** is the obtaining of property through the threat of future harm. It's also called blackmail.

In some states, the crime of extortion occurs once the defendant has made the request for property accompanied by the threat of future harm. In other states, the victim actually has to hand over property for it to be extortion (if the threat is made but no property turned over, it could be attempted extortion).

The threat of future harm does not have to be physical harm—it can be any type of harm: monetary, reputational, or otherwise.

The threat of harm also doesn't have to be to the victim. It can be to anyone.

When Your Killer Represents Your Mistress in Her Divorce from Your Blackmailer

You have undoubtedly heard of Alexander Hamilton. He is mainly famous for four things: [62] He helped establish the U.S. Constitution, he wrote the Federalist Papers, he is the subject of the Broadway play bearing his name, and he was the subject of the trivia question in the first "Got Milk?" commercial.[63]

Hamilton was also involved in the nation's first big sex scandal—an affair that made Hamilton the target of an extortion scheme.

Maria Reynolds, a married woman, approached a married Alexander Hamilton and asked if he would help her financially because her husband had abandoned her. Hamilton, eager to help, did not have any money on him at the time, so he procured the address of the place she was staying to deliver the money to her later.

When Hamilton arrived, she invited him into her bedroom where, according to his account, "some conversation ensued from which it was quickly apparent that other than pecuniary consolation would be acceptable." [64] Ever the

---

[62] He was also the founder of the Federalist Party the United States Coast Guard, the U.S. financial system, and *The New York Post*.
[63] http://y2u.be/OLSsswr6z9Y (last accessed July 31, 2018).
[64] That's colonist English for, "she wanted me for more than just my money."

philanthropist, Hamilton was willing to help the poor damsel in any way he could.

Maria's husband, James Reynolds, who had apparently not abandoned her, knew about the affair almost immediately. Knowing how much Hamilton had to lose if the affair were known, he invited Hamilton to pay him to keep the affair quiet. In fact, for the duration of the one-year affair, James Reynolds encouraged Hamilton to keep seeing his wife "as a friend," and at the time of each "friendly" rendezvous, Reynolds would extort a new payment from Hamilton. He collected a total of $1,300.

When the affair ended, so did the extortion payments, though the truth of the affair came out several years later when Hamilton's known nemesis, Thomas Jefferson, released the news, severely damaging Hamilton's reputation.

In an interesting twist of fate, when Maria Reynolds eventually petitioned to divorce her husband, James, her divorce attorney was none other than Aaron Burr—the man who would famously and fatally wound Alexander Hamilton in a duel eleven years later.

Perhaps even more bizarre, when Aaron Burr's wife divorced him after four short months of marriage—29 years after the duel—her divorce attorney was Alexander Hamilton's son and namesake, Alexander Hamilton Jr., who, by taking away Burr's wife from him, found a satisfying and legal way to avenge his father's death.

RECEIPT OF STOLEN PROPERTY

Receipt of stolen property is unlike the prior crimes in that the defendant does not actually do the stealing. To be guilty of **receipt of stolen property**, the defendant must have possessory control of property that she knows another has stolen and must also intend to deprive the owner of the property.

Actual knowledge that the property is stolen is not always necessary. Constructive knowledge can be imputed to the possessor based on circumstantial evidence, like if you buy it from a seller with a reputation for selling stolen goods, or who is acting suspiciously when you buy it, or who sells it at a price that's too good to be true.

Actual possession of property known to be stolen is not enough. You must also intend to deprive the true owner of it. If you acquire property you know is stolen for the purpose of returning it to the true owner, you have not committed the crime of receipt of stolen property.

eBay Fencing Scheme

The key to making money in the retail business is to sell products for more than you pay for them. The more you sell and the bigger your margins, the more you make.

Robert A. Hill had a very successful eBay store, atlantis_discount_warehouse_llc, which he operated out of Roswell, Georgia. In a decade's worth of sales, Hill made over $9 million. The secret to his success? Go into business with a team of identity stealers who would purchase electronics using their victims' credit cards. Pay them 60% of the retail value of the goods, then mark it up, sell it on eBay as new, and watch the money roll in.

Anyone who has ever started a business understands that where there is no risk, there is no reward. But the savviest business owners don't take the types of risks that could get them seven-plus years in federal prison for receipt of stolen property, which is what Hill got when he was caught.

FORGERY AND UTTERING FALSE INSTRUMENTS

Forgery itself is not a form of theft, though it can be an instrument for theft. Creating or altering a false instrument with the intent to defraud is **forgery**.

The crime occurs once the instrument is created or altered. It doesn't need to be used, and it does not have to trick anyone into believing it is real.

It is also not forgery if the writing is authentic but contains a false statement. Adding a zero to a check is not forgery because the check is authentic. That would be false pretenses. But if you sign someone else's name to the check, it changes the identity of the instrument and is now a forgery.

A related crime, **uttering a false instrument**, involves putting the forged instrument into circulation, trying to pass it off as genuine, with the intent to defraud.

The *mens rea* of forgery and uttering a false instrument is the same: intent to defraud. But the *actus reus* is different. For forgery, it is the act of creating or altering the false document. For uttering a false instrument, it is offering the false document as genuine.

Did you follow all that? You may have to read it a few times. I know I did.

Brokering a Record Deal
Crawley, Texas' Charles Ray Fuller has a personal motto: "go big or go home." Charles had big plans to start a record company. A really big record company. At 21-years-old, he had not yet saved up the $360,000,000,000 (that's 360 billion

dollars) in start-up capital that he calculated he would need.[65] But he was nothing if not determined and had a plan for coming up with the money fast: He "borrowed" a blank check from his girlfriend's mom's checkbook, wrote out a check for $360,000,000,000 which, if you were wondering, would be about 3,600 *crates* of $100 bills (enough that you'd need a fleet of moving trucks to move it), and signed his girlfriend's mother's name to it (forgery).

When Fuller went to cash it (uttering a false instrument), he ran into three problems: (1) the bank did not have that kind of cash on hand and would not have given it to him if it did, (2) his girlfriend's mother did not have that kind of cash in her bank account, and her daughter would not have been dating him if it did, and (3) the check he forged was not written out to him. Oops. That's what we call the dumb-criminal hat trick.

Were you hoping for a convenient chart that laid out important attributes of the different property crimes? No? Well, then this is not the chart you're looking for. There's nothing to see here. Move along.

Crime	How Property Is Obtained	Other Attributes
Larceny	Trespassory taking	Carrying away
Embezzlement	Property is entrusted	Property can be given away or destroyed
False Pretenses	Fraud	Title passes

---

[65] To put that in perspective, the biggest record company in the world, Universal Music Group, is worth about $40 billion. Three-hundred and sixty billion is so far above the mark that even Dr. Evil would laugh.

Crime	How Property Is Obtained	Other Attributes
Larceny by Trick	Fraud	Title does not pass
Robbery	Force or threat of immediate harm	Taken from the person or in their presence
Extortion	Threat of future harm	Harm can be reputational
Receipt of Stolen Property	From someone other than the owner	You must have intent to deprive the owner
Forgery/ Uttering	False instrument	No need to obtain property

CHAPTER EIGHT

Offenses Against the Habitation

WHEREAS GENERAL PROPERTY CRIMES protect our property rights—including the rights of control, enjoyment, disposition, exclusion, and possession—offenses against the habitation are more about the protection of our security and privacy.

Home as a place of comfort and security is something built into society's collective psyche, reflected by common aphorisms about home, like "home is where the heart is," "there's no place like home," and "home sweet home."

When our place of comfort and security is invaded or taken from us, there is a real sense of loss and fear. Imposing punishments for crimes against habitation is society's way of recognizing the value of a home while attempting to protect it.

BURGLARY

Do you like the word *burgle* much as I do? It isn't used near as much as *burglary* is. Even *burglarize*, a synonym for *burgle*, just doesn't get much use.

Did you know that's something you can actually verify? With Google's ngram book viewer, you can actually see how common words are used in print for the last two centuries:

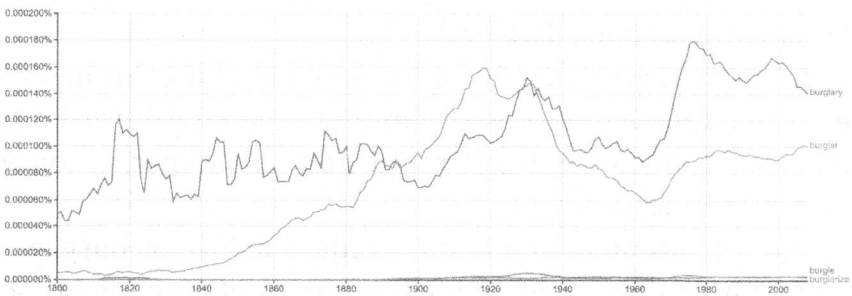

You can see, for example, that the verbs *burgle* and *burglarize* never really took off, and that the nouns *burglary* and *burglar* have always been preferred, finding peak use, for some reason, in 1975 and 1929, respectively.

Burglary as a crime is much more interesting than as a word, though. Originally, burglary had some very distinctive elements. **Burglary** was breaking and entering into someone else's dwelling, at night, with the intent to commit a felony.

Breaking and entering, itself a lesser crime (does not require felonious intent), requires trespassory entry by minimal force. To constitute "breaking," pushing open a partially open or unlocked door is enough if there is no permission to be there. It is also "breaking" if consent to enter (or the opening of the door) is procured by fraud.

"Entering" is also broadly interpreted. Shooting bullets into a house constitutes entry. Sticking an arm through a window to unlock the door is also entry. In one case, a man left his finger behind when a closed window removed it from his hand. That was also entry. And gross.

The term "dwelling" in burglary laws is also broadly construed. It doesn't actually have to be a house. It is enough that someone uses it regularly to sleep in, even if they aren't in it at the time of entering, or it is used also as a business, or it has wheels (like an RV). On the other hand, a house that is being built but is not yet occupied, or one that has been abandoned permanently, is not a dwelling.

It is occupancy of the dwelling, not ownership, that determines whether burglary is committed. A landlord breaking into a house she owns, but is renting out to another, is still a burglar. (A burglord?)

One of the more interesting requirements of burglary was that it occur at night. When it came to defining "night," it was not a simple task of marking off those hours between dusk and dawn.

Remember how I told you that lawyers like to use Latin words to make us feel smarter? Consider these definitions of "night" for purposes of determining whether a burglary had occurred:

> "If the sun be set, yet if the countenance of a party can be reasonably discerned by the light of the sun or crepusculum [Latin for twilight], it is not night nor noctanter [Latin for night] to make a burglary."

> "As to what reckoned night, and what day, to this purpose: anciently the day was accounted to begin only at sunrising, and to end immediately upon sunset; but the better opinion seems to be that if there be daylight or crepusculum enough begun or left to discern a man's face withal, it is no burglary."

So, there you have it. At one point, if there was enough light left to recognize a man from his face, he's not a burglar. It's only burglary if it's too dark to tell who he is. It seems there would be some practical limitations to that rule that would make it very difficult to prosecute someone for burglary. And there undoubtedly was at least one enterprising would-be burglar who, after being caught, offered the following creative defense: "Your honor, the fact that I am here before you should be proof enough that I am no burglar. If the police recognized my face enough to arrest me, my actions don't amount to burglary. If I truly were a burglar, they never would have known who I am, and I would not be here."

The last requirement—that the entry occur with the intent to commit a felony, is just that. The felony doesn't have to occur or be successful. Just the fact that it was intended is enough.

Modern burglary statutes have further broadened the crime. It is no longer necessary that any "breaking" occur. Entry is enough.

In some cases, being invited in but staying beyond when your invitation expires, like hiding in a store until after it is closed and everyone is gone, acts as a substitute for entry, called **surreptitious remaining**.

The dwelling requirement in many states includes the yard around the dwelling, or even a vehicle, whether or not in close proximity to a dwelling.

A burglary also does not have to occur at night anymore, and even an intent to commit a misdemeanor, rather than a felony, qualifies.

You can probably guess why these rules have been broadened. As the crime was originally defined, these would all be defenses to burglary:

- I slipped through an open door without touching it (no breaking);
- They invited me in; they just didn't realize I never left (no trespassory entry);
- I only took stuff from the backyard (not a dwelling);
- The victims discerned my face by the light of the waning sun (not night); and
- I only took $5 (a misdemeanor, not a felony).

Now we're going to call it a burglary if someone enters almost any of our private property for almost any improper purpose.

Goldilocks Burglar

A British couple in 2014 returned home after a vacation to find a Polish man, Lukasz Chojnowski, napping in their bed. He had washed their dishes, laundered his underwear, and even stocked their kitchen with groceries. Apparently, he felt their home fit him "just right."

Criminal Addiction to Facebook

In 2009, Jonathan G. Parker burgled a home in West Virginia. But even amidst the break-in, he just couldn't make the burglary his highest priority. Apparently, he had a bad case of FOMO[66] because he logged into Facebook using the homeowner's computer to find out what hilarious memes had been posted in his absence—something he needed to know before he finished his felony.

His addiction sated for the time being, Parker left with two diamond rings. He did not take the computer, and he did not log off his account. The homeowner got her rings back, and Parker got a new roommate. I hear he's getting lots of

---

[66] Fear of missing out.

"likes" in prison.

ARSON

Like burglary, arson is a crime against a dwelling, which crime has broadened in recent years. **Arson** is the malicious burning of someone else's dwelling.

The meaning of dwelling for arson is the same as it is for burglary, though there is rarely any requirement that it ever be used for the purpose of sleeping. A shed, for example, can be the subject of arson. A forest can, too.

Burning from fire is actually required and constitutes the *actus reus* element. Discoloration from smoke, or mere charring, is not enough. At one time, damage from an explosion did not constitute arson, but in most jurisdictions, it now does.

Malice, the *mens rea* of arson, does not require any particular reason or motive for the burning. It is enough that the one lighting the match (or mixing combustible chemicals or getting creative with a magnifying glass) have at least reckless disregard for setting a structure on fire. Knowledge that it will burn or intent that it burn is obviously enough.

Who's the Fairest Arsonist of All?

Just because you're a pyromaniac doesn't mean you shouldn't look your best.

Ohio resident Donald "Chip" Pugh was wanted in connection with several crimes, including arson. The Lima police took to social media and posted a mugshot taken of him from an earlier arrest, asking for the public's help to locate him.

Pugh was made aware of the mugshot, which he didn't feel showed his best side, so he put product in his hair, trimmed his mustache, put on some sunglasses and a tweed suit, and took a photo of him driving. He sent that to police with the caption, "Here is a better photo that one is terrible [sic]."

His response raised awareness even more, and he even explained himself on a local radio station: "Man, they just did me wrong. They put a picture out that made me look like I was a Thundercat...[67] or James Brown on the run. I can't do that."

News of his concern over his own image reached Florida, where he was arrested. This time when they took his picture for the booking, he smiled big. He wasn't about to repeat the ugly-mugshot fiasco that led to his arrest.

---

[67] Pugh does resemble James Hyman, who voiced Panthro in the original Thundercat series.

CHAPTER NINE

Offenses Involving Judicial Procedure

CRIMES AGAINST THE PERSON and against property directly address the wrongs that we, as a society, want to prevent. Offenses involving judicial procedure, on the other hand, are secondary protections—put in place to make sure there are additional consequences for both (1) those who unrighteously seek to evade the consequences of a crime they've committed and (2) those who help them.

They say the cover-up is worse than the crime. Perjury, subornation of perjury, and bribery exist to make sure the cover-up is also a crime.

PERJURY

In most of the modern world, there is a presumption of innocence. The burden is on the accusers to bring forth enough evidence to convict the accused of the crime. That evidence can take many forms, including testimony of witnesses. Because the stakes are potentially so high (loss of freedom or life), and to protect the integrity of the judicial system, we impose consequences on those who testify and bear false witness.

This is no less true in civil cases, where instead of freedom or life hanging in the balance, it could be someone's livelihood, their home, or compensation for something lost.

Warren E. Burger, once Chief Justice of the United States Supreme Court, said this of perjury:

> *In this constitutional process of securing a witness' testimony, perjury simply has no place whatever. Perjured testimony is an obvious and flagrant affront to the basic concepts of judicial proceedings. Effective restraints against this type of egregious offense are therefore imperative. The power of subpoena, broad as it is, and the power of contempt for refusing to answer, drastic as that is and even the solemnity of the oath cannot insure [sic] truthful answers. Hence, Congress has made the giving of false answers a criminal act punishable by severe penalties; in no other way can criminal conduct be flushed into the open where the law can deal with it.*[68]

Perjury occurs when a witness to a judicial proceeding takes an oath to tell the truth and then willfully lies about a

---

[68] *United States v. Mandujano*, 425 U.S. 564, 576, 96 S. Ct. 1768, 1776, 48 L. Ed. 2d 212 (1976).

material issue.

The lie must be material to be perjury, meaning it must be such that it could affect the outcome. And the lie must be willful, meaning there is actually an intent to deceive. It is not enough to be mistaken, and if you honestly believe what you're saying is true, it's not perjury. Additionally, to be a lie, it must be a false *factual* statement; opinions cannot be lies.

Bill Clinton's Perjury Case

In a prime example of the cover-up being worse than the crime, as the sitting president of the United States, Bill Clinton defended himself in a civil lawsuit for alleged improper sexual conduct with Paula Jones years before when he was the governor of Arkansas.

Bill Clinton responded to several questions, under oath, during a deposition, for which he later faced charges of perjury. Three of the questions (paraphrased), along with the exact answers, and Bill Clinton's defense to the perjury charges, are the following:

Q: Did you have a "sexual affair" with Monica Lewinsky?
A: No.
Defense: Bill Clinton believed a sexual affair to equate to sexual intercourse, and he never had sexual intercourse with Monica Lewinsky, so he believed he was telling the truth, which is not perjury.

Q: Here is our definition of "sexual affair": "contact with the genitalia, anus, groin, breast, inner thigh, or buttocks of a person with an intent to arouse or gratify the sexual desire of any person." Using that definition, did you have a "sexual affair" with Monica Lewinsky?
A: No.
Defense: Perjury cannot be proved with "an oath against

an oath," so if they want to prove that this is false, they need more than just testimony from Monica Lewinsky.

Q: Were you ever alone in a room with Monica Lewinsky?
A: I don't recall.

Defense: He never said he was never in a room with her. He said he did not recall. And there is no proof that he actually did recall.

Bill Clinton was eventually acquitted of the perjury charges, proving that he deserves the nickname, "Slick Willy."[69]

SUBORNATION OF PERJURY

Perjury is a crime that punishes the person who gives the false oath. Subornation of perjury, on the other hand, punishes not the one who speaks the lie, but the one at whose request the lie is told.

"To suborn" means to bribe or induce someone to do something unlawful. **Subornation of perjury** occurs when one person induces another to commit perjury and the other actually commits perjury.

This is a crime that sends the message that the one telling a lie in a judicial proceeding is not the only one taking a risk. Those encouraging the crime have also committed a crime.

Bill Clinton was also accused of suborning perjury by allegedly trying to convince both Monica Lewinsky and Linda Tripp to lie under oath about their relationship with him.

---

[69] He did, however, settle with Paula Jones, paying her $850,000 to dismiss the lawsuit against him.

BRIBERY

Providing false testimony, or perjury, is one way to try to rig the justice system. So, too, is **bribery**, which is the transfer of something of value in exchange for official action.

Both the giver and the receiver can be guilty of bribery. A bribe does not have to be monetary. It can something else of value, like a favor.

Traditionally, it was only judges who could be subject to penalties for accepting bribes. Now with broader applicability, bribery applies to anyone in an official capacity, including jurors, police officers, court officials, border patrol agents, etc. It even extends to the private sector. Paying an athlete to throw a game is bribery.

For Sale: One Seat in the U.S. Senate

Barack Obama was one of Illinois' representative in the United States Senate while he was running for president in 2008—an election he won.[70] Then-governor Rod Blagojevich had the unilateral authority to appoint Barack Obama's successor, who would fill the vacancy between when Barack Obama gave up his seat and when the Illinois electorate selected his replacement.

Unknown to Governor Blagojevich, at the time, he was under investigation by the FBI, which was gathering evidence for corruption charges against him.

After his December 2008 arrest, Blagojevich was charged with, among other things, solicitation of bribes and conspiracy to solicit and accept bribes. Blagojevich viewed the senate seat as "a [expletive] valuable thing, you just don't give it away for nothing," as an FBI wire-tapped recording revealed.

---

[70] I hope you knew that.

It took two trials (the first one ended in a mistrial due to a hung jury), but a jury eventually convicted Blagojevich of two inchoate crimes related to bribery (attempt and conspiracy), among other crimes (17 counts total). He was sentenced to 14 years. He is serving in a federal prison in Colorado and expected to be released in May 2024. I bet he's wishing now he gave that seat away for nothing.

PART 2: CRIMINAL PROCEDURE

CHAPTER TEN

Introduction to Criminal Procedure

CRIMINAL LAW COMPRISES A SET of rules that a federal, state, or local government establishes to govern the behavior of everyone within its boundaries by imposing punishment for violations of these rules. Its purpose is to promote safety, security, and privacy within the jurisdiction of the sovereign entity.

Criminal procedure, on the other hand, is the set of rules that applies *to the government*—law enforcement, prosecutors, and judges. Procedural rules are designed to protect the accused and the integrity of the criminal justice system.

Many of the limits on police power are found in the

Amendments to the United States Constitution, though when the Framers drafted the Constitution, law enforcement was far different then than it is now. Police officers were not organized into departments until the 19th century, nor did they typically investigate the crime. As police began to band together into departments, they had plenary authority to apprehend and charge criminals with little oversight.

This led to a highly powerful police force that eventually evolved into the hotbed of police corruption that dominated during Prohibition in the 1920s. As the public demanded police oversight, the judicial branch took on that role and, with the authority of constitutional amendments, placed limits on police power.

These limits exist today in many forms, predominantly as rules of criminal procedure.

CHAPTER ELEVEN

The Fourth Amendment

DURING AMERICA'S DAYS AS A colony to England, King George II issued writs of assistance that allowed the holder of the writ unfettered access to the property of the colonists. These writs would not expire as long as the king lived, were transferable, and provided immunity to the holder for any property damage resulting from a search. Those holding the writs of assistance were above the law, leaving the colonists victims in their own homes.

When George II died in 1760, a six-month countdown until their expiration commenced. This emboldened colonists, who began challenging the legality of these writs even before they expired. Merchant Daniel Malcolm refused to allow customs officials to search his cellar even though they had a writ of assistance. He argued that such a writ was unlawful and had the support of dozens, if not hundreds (reports vary) of colonists gathered at his house.

Because of the uncertainty of the legality of these writs, British Parliament passed the Townsend Act in 1767 confirming that they were, in fact, legal, but this did not appease colonists. John Hancock famously refused to allow a search of his ship, the *Lydia*, the next year in 1768.

These conditions precipitated the American Revolution and prompted the colonists, after having defeated Britain in the war for their independence, to draft into the Constitution, via amendment, protections against unreasonable searches and seizures.

Amendment 4: The right of the people to be secure in their persons, houses, papers, and effects, against unreasonable searches and seizures, shall not be violated....

The Fourth Amendment, part of the original Bill of Rights, was enacted in 1791. Its application is broad, but not unlimited. It specifically protects against unreasonable searches and seizures.

The Fourth Amendment only acts to limit searches and seizures that the government performs or authorizes. Although the Fourth Amendment originally applied only to the federal government, the Fourteenth Amendment, passed in 1868, expanded its application to state governments as well.

The Fourth Amendment now applies not just to federal officers, but also to public school teachers, staff of state hospitals, social workers, and tax appraisers. It does not apply to actions of a private individual.

SEARCHES

A **search** is an intrusion by a law-enforcement officer (or other governmental officials) into an area reasonably expected to be private. A search is only reasonable if it is in an area where (1) there is no reasonable expectation of privacy, (2) the government actor has a warrant, or (3) there is a valid warrant exception (warrant requirements are also discussed later in this chapter).

This interdiction against unreasonable searches and seizures is limited to areas where a person has a **reasonable expectation of privacy**. That means (1) they are actually subjectively treating it as private, and (2) the subjective expectation of privacy is one society is prepared to accept as reasonable.

Searches of the Home

A search of the inside of someone's home can never legally occur without a warrant or warrant exception as a home is a place of utmost privacy.

A police officer may approach the front door of your home, just as anyone else in the public may, and anything visible to him from that vantage point can be used as evidence, but a dog sniff at the front door of your home is a search and requires either a warrant or a warrant exception.

What about areas outside the home? What can be searched any time and what requires a warrant? Courts draw the distinction between curtilage and open fields. Open fields can always be searched, whereas searches of the curtilage require a warrant or warrant exception.

Curtilage is the area immediately surrounding the home and outbuildings near the home that are intimately tied to the use of the home. To determine whether an area falls within the definition of curtilage, and is therefore protected by the Fourth Amendment, courts will look at four factors: (1) how close is the subject area to the home? (2) is it fenced in or otherwise in an enclosure with the home? (3) how is the area used? And (4) what has the owner done to keep the area hidden from view from passersby?

Curtilage protections do not extend to evidence viewed from any place where a member of the general public could legally view it. Thus, if a police officer flies over a house and is not in violation of any airspace restrictions, this would be a valid search, even without a warrant. The same would undoubtedly apply to evidence captured by drones or images on Google Earth—as long as it is obtained in a manner that someone in the public could obtain it. This principle is called the **fly-over exception**.

Open fields, by comparison, include all the owner's property that is not curtilage and fall outside the scope of Fourth Amendment protections.

If the person asserting the privacy right does not live in the home being searched, their privacy rights are further limited. If they are an overnight guest, the protections still apply as though they live in the home. If, on the other hand, they are only there for a brief time, and are conducting business instead of for social purposes, they have far fewer privacy rights, and warrantless searches are more likely to be validated as used against them.

As technology advances, constitutional questions arise that the Founding Fathers could not have contemplated. For example, in 2001, courts had to determine whether the use of thermal imaging technology to read higher-than-normal heat signatures in a man's garage—which police believed indicated the use of grow lights to cultivate marijuana—constituted a search of a private area under the Fourth Amendment.

After reading levels of high heat, police obtained a warrant and found over 100 marijuana plants growing in the defendant's home. The United States Supreme Court, in a divided decision, concluded that use of this technology did constitute a search that unreasonably invaded the defendant's privacy even though it was conducted from outside the home. The marijuana police found was therefore disallowed as evidence against the defendant.[71]

Searches of the Person, Papers, and Effects

The Fourth Amendment protects against unreasonable searches of "persons, houses, papers, and effects." We have discussed how this Amendment has been interpreted in regard to "houses." It also applies to vehicles, which were not specifically contemplated by the Framers.

Only private areas of vehicles are protected. A public actor may examine the outside of a vehicle and look through windows—just like any other person may do. Typically, a search of the interior of the vehicle, like under the seats, in the trunk, or in the glove compartment, require a warrant or warrant exception, though for some reason police are allowed to reach into an open window to move papers to see the vehicle identification number. But attaching a GPS to a vehicle is not allowed—at least without a warrant or warrant exception.

What about dog sniffs? Are those searches under the meaning of the Fourth Amendment? It depends. If a dog sniffs a car during a traffic stop, where the police have lawfully

---

[71] *See* The Exclusionary Rule in Chapter 12.

pulled the vehicle over, and they don't keep the occupants any longer because of the dog's sniffing than they would otherwise, this is not an unconstitutional search.

Items worn, carried, and stored, like a wallet, jacket, and purse, are personal effects that can only be viewed externally without a warrant or warrant exception. Luggage, too. Police can look at your luggage, but to open it, or even feel it for lumps or bulges, normally requires a warrant. But a dog sniff of luggage is allowed.

Police may always search **abandoned property** without a warrant, including luggage discarded while fleeing from police, a house that no one lives in, and garbage left out for collection.

New technology also creates interesting questions under the Fourth Amendment when it comes to searches of computer files and email.

Middle School Strip Search

Savana Redding was a thirteen-year-old girl attending Safford Middle School in Arizona. She was an honor-roll student and had no disciplinary history. Nevertheless, when classmate Marissa Glines was caught with prescription-strength Ibuprofen, Marissa told the principal that Savana gave the pain medication to her.

Without any further questions about Savana's role, Principal Wilson called Savana to his office. She denied ever having or providing the pills to other students and consented to a search of her backpack, which turned up nothing.

Principal Wilson believed that students kept contraband in their underclothing, so, without ever contacting her parents, he sent Savana to the nurse's office, where she was told to strip down to her underwear, "pull her bra out and to the side and shake it," and "pull out the elastic on her underpants" to see if that would shake anything loose. It didn't. No contraband was found.

Savana's mother filed a lawsuit on her behalf, which eventually made its way to the U.S. Supreme Court. For police to conduct such a search, the Court reminded us, requires either a warrant or "probable cause." However, school officials (who are also government actors), merely require "reasonable suspicion." But there must also be a proportional relationship between the invasion of privacy and the seriousness of the suspected infraction.

Ultimately, a search of her outer clothing and backpack was justified, but because the alleged infraction (distribution of Ibuprofen) was not serious enough to justify the indignity of a strip search, Savana's Fourth Amendment rights were violated.

Mass High School Police Groping

When an illegal search occurs, the police cannot use any evidence found by the illegal search against the defendant.[72] But what if the police don't find any evidence? What are the consequences then? The Worth County Sheriff's department in Georgia paid a steep price to find out.

Sheriff Jeff Hobby, acting on information that thirteen students at Worth County High School possessed illegal drugs, went with about 40 other officers to the school on the morning of April 14, 2017, and put it on lockdown. For the next four hours, he and his posse of state officials searched virtually every student—over 800 in total—including those with disabilities, which searches allegedly included groping of genitals of both male and female students. The police found no drugs.

---

[72] *See* Exclusionary Rule in Chapter 12.

Besides being the subject of a class-action complaint that resulted in a $3 million settlement (the limits of the sheriff's insurance policy), Sheriff Hobby was suspended, and he and two of his deputies were criminally indicted for their role in the illegal searches.

SEIZURES, ARRESTS, AND DETENTION

Seizures, like searches, are defined broadly. A **seizure** occurs when a government representative exercises control or takes possession of a person or property.

The Fourth Amendment does not apply to all seizures: it merely protects people against *unreasonable* seizures. To be reasonable, a seizure must be conducted in accordance with a warrant or meet one of the warrant exceptions.

Seizures and Police Stops

An arrest is a seizure of the person and must comply with the Fourth Amendment. An **arrest** occurs, even if the police have not handcuffed you and put you in the back of a patrol car or taken you into custody (which would very obviously be an arrest), when in the totality of the circumstances, a reasonable person in your position would not feel free to decline the officer's requests or terminate the encounter.

An arrest is distinguished from an **investigatory stop**, which is also a seizure of the person, but where under the totality of the circumstances, it does not amount to an arrest. When the police pull someone over who is driving, this is an investigatory stop of the driver and the other vehicle's occupants, and such a stop can only be done within the limits set by the Fourth Amendment.

If an officer merely approaches someone and asks a few questions, but there is no detention, it is not an arrest, not an investigatory stop, and therefore not a seizure and not subject to Fourth Amendment protections.

Police activity is only a seizure if there is some show of force or other authority.

Arrests, which are always seizures, are very common. The FBI estimates that more than 250 million arrests occurred in the 20 years ending in 2014. That's one arrest every 2.5 seconds.[73] To put that in perspective, a new baby is born in America approximately every 8 seconds, so for every new birth, there are 3.2 arrests.[74]

WARRANT REQUIREMENTS

Amendment 4: ... and no warrants shall issue, but upon probable cause, supported by oath or affirmation, and particularly describing the place to be searched, and the persons or things to be seized.

As we have explored the limits of reasonableness for searches and seizures under the Fourth Amendment, we have frequently distinguished between those searches and seizures that require either a warrant or warrant exception and those that do not. If the search or seizure is the kind that falls within Fourth Amendment protections, it must be accompanied by a warrant or fall under a valid exception to

---

[73] That's the same rate that Nutella is being purchased. So, every time you see an arrest, know that another bottle of Nutella has found a home.

[74] There are approximately 12.5 million arrests every year and almost 4 million births.

the warrant requirement.

A **warrant** is a legal document issued by a government official that authorizes a government actor to carry out some action related to the administration of justice. Usually, a judge issues the warrant, and it allows the police to conduct a search (search warrant) or make an arrest (arrest warrant). The ultimate purpose of a warrant is a seizure—the seizure of evidence found when executing a search warrant, or the seizure of the person subject to the arrest warrant.

For a warrant to be valid, there are three requirements:

1. Neutrality;
2. Probable Cause; and
3. Particularity.

Neutrality

When police want to search an area where there is a reasonable expectation of privacy, or arrest someone, and there is no valid warrant exception, they must apply for a warrant. The official they petition for a warrant, and who has authority to issue a warrant, must be **neutral** and detached—it cannot be an official associated with law enforcement, like the attorney general; or an official with ties to the case, like a judge who is the victim of the crime or a judge who gets paid for every warrant she issues.

The neutrality requirement relates specifically to the official issuing the warrant and is designed to assure police compliance with constitutional standards.

In a 1977 United States Supreme Court case out of Georgia, the High Court had to examine the neutrality of a justice of the peace who issued a search warrant to search a home that resulted in the discovery and seizure of illegal marijuana.

At the time, a justice of the peace was not required to be

a lawyer, have gone to law school, or have any other legal training. Salaries for Georgia justices of the peace were governed by statute, which provided that they would be paid $5 per warrant issued, and $0 for a warrant application reviewed but denied. The justice in this particular case had, in his history as a justice, issued some 10,000 warrants (and presumably been paid $50,000 as a result).

The Supreme Court concluded that this justice of the peace was not neutral as there was a built-in incentive for him to issue a warrant without a valid probable-cause analysis. Lacking the neutrality required of warrants, the warrant that resulted in the discovery of marijuana was invalidated, and the defendant's conviction reversed.[75]

Probable Cause

The Fourth Amendment requires that an application for a warrant be supported by probable cause, as set forth "by oath or affirmation." This is typically done by affidavit in the warrant application. An **affidavit** is a written statement, made under oath. It is a sworn statement.

In the affidavit accompanying the warrant application, the officer seeking the warrant lays out what they believe is probable cause for the issuance of a warrant. The magistrate then must evaluate whether probable cause exists in deciding whether to issue the warrant.

Probable cause is "[a] reasonable ground to suspect that a person has committed or is committing a crime [for an arrest warrant] or that a place contains specific items connected with a crime [for a search warrant]."[76]

To support probable cause for a search warrant, the

---

[75] Also, Georgia had to figure out a new way to pay justices of the peace.
[76] Black's Law Dictionary, "Probable Cause," (10th ed. 2014).

affidavit must convince the issuing official that (1) the items subject to the search warrant are connected to a crime, and (2) that when law enforcement executes on the warrant, they will find those items in the place identified on the warrant.

To support probable cause for an arrest warrant, the affidavit must convince the issuing official that (1) a specified criminal offense was committed (2) by the person subject to the arrest warrant.

Particularity

The third and final warrant requirement is particularity. To be valid, a warrant (or documents incorporated by reference by the warrant) must **particularly** describe the items sought to be seized and the places to be searched. The particularity requirement "makes general searches under them impossible and prevents the seizure of one thing under a warrant describing another. As to what is to be taken, nothing is left to the discretion of the officer executing the warrant."[77]

This rule does not prevent officers executing a warrant from seizing contraband or other evidence they find somewhere other than where they've indicated in the warrant if that item is in open view while they are otherwise properly conducting a search.

Wiretapping

One type of search that is of considerable public interest is government **wiretapping**, which is the electronic eavesdropping of telephonic communications. Wiretapping constitutes an intrusion into and violation of our constitutionally protected reasonable expectation of privacy. Thus, for the government to legally (and constitutionally) tap into

---

[77] *Marron v. United States*, 275 U.S. 192, 196, 48 S. Ct. 74, 76, 72 L. Ed. 231 (1927).

our private conversations, they must have a warrant.

Because wiretapping represents a significant level of intrusion, it has its own special set of requirements to qualify for a warrant. In each warrant application, the applying officer must identify him- or herself and provide the following:

1) Details as to the offense being investigated,
2) A particular description of the facilities the communication is to be intercepted from;
3) A particular description of the type of communications sought to be intercepted;
4) The identity of the person, if known, committing the offense and whose communications are to be intercepted;
5) An explanation as to what other less-intrusive measures have been attempted and failed or a reason they are not likely to work or are too dangerous to try;
6) A description as to how long the wiretap would last; and
7) Any outcome of previous wiretaps or attempts to get a wiretap for the same individual.[78]

The federal government and many states do not require a warrant to intercept private telephonic conversations where one party to the conversation consents to the recording. For example, in those cases where a felon is conversing with a government informer who is on the call knowing the conversation is being recorded, no warrant is necessary.

Other Warrant Limits

Once the magistrate has issued the warrant, its execution

---

[78] *See* 18 U.S.C. § 2518 for a full description of the federal warrant wiretapping requirements.

must be reasonable to comply with the Fourth Amendment. Thus, it can only be executed by a police officer, and only within certain times. The rules impose an expiration date on the warrant (10 days in federal court), the warrant itself will have an expiration date, and absent justified circumstances, the warrant must be executed during waking hours, which vary state to state, but it will be something like 6:00 a.m. to 10:00 p.m.

The scope of the search must also be limited to what the warrant allows, including the locations described in the warrant and those places reasonably necessary to discover the items the warrant describes. It does not allow the search of people not described in the warrant, for example, though if they can otherwise validly arrest someone present during an authorized search, the police can subsequently search that person incident to the arrest.[79]

Police, can, however, detain persons during a search to promote officer safety and make the search more orderly or to prevent escape.

The process for executing a warrant that has evolved through judicial decisions requires the police to "knock and announce," meaning they must knock on the door, announce themselves and their purpose ("this is the police; we have a warrant for your arrest"), and then give the occupant a reasonable amount of time to open the door. Fifteen to twenty seconds is enough. Police have apparently had too many experiences where the person inside shouts, "hold on, I'm in the shower, let me get dressed" while they frantically flush drugs down the toilet.

If the attendant circumstances are such that there is a reasonable suspicion of substantial risk that knocking and

---

[79] This warrant exception is discussed in more depth on page 107.

announcing would be dangerous or futile (including officer danger or where it is not necessary because the occupants already know the police are there) or would inhibit the investigation (would result in destruction of evidence or escape), police can enter without knocking or announcing their presence.

After executing the search, police fill out a form indicating that the search warrant was executed and provides information about the time and place of its execution. Anything the police seize during a search must be listed on an inventory sheet and attached to this form. Then the owners of the seized material must be made aware of what exactly was taken so they have an opportunity to challenge the validity of the seizure.

In a 1999 United States Supreme Court case, the nine justices had to decide whether a media "ride-along," while police attempted to execute an arrest warrant, was a violation of the Fourth Amendment.

Police in that case, arising out of Maryland, obtained an arrest warrant, and early one morning, went to arrest a man in his home. They arranged for a newspaper reporter and photographers to accompany them. The reporter and photographer documented the sweep of the home. The subject of the arrest warrant was not there, though his parents were and subsequently sued.

The parents challenged the warrant's execution, and the Supreme Court unanimously agreed, "it violates the Fourth Amendment for police to bring members of the media or other third parties into a private dwelling during the execution of a warrant unless the homeowner has consented or the presence of the third parties is in aid of the execution of the warrant."[80] The police were nonetheless released from the case based on qualified immunity that allowed their actions because there had been no similar case decided and the rule was not clear before they had the ride-along.

EXCEPTIONS TO THE WARRANT REQUIREMENT

Under the Fourth Amendment, it is generally preferred, for the protection of our rights to "persons, houses, papers, and effects," that searches and seizures only be incident to the issuance of a warrant by a neutral magistrate, which is based on probable cause.

However, there are some exigent circumstances where it could promote crime or be dangerous to obtain a warrant prior to a search or seizure. In those cases, a warrantless search is justified and allowed within Fourth Amendment jurisprudence.

We will discuss several circumstances where warrantless searches and seizures do not run afoul of Fourth Amendment protections:

1. arrests;
2. searches incident to an arrest;
3. motor vehicle searches;
4. plain view;
5. consent;
6. stop and frisk;
7. hot pursuit;

---

[80] *Wilson v. Layne*, 526 U.S. 603, 618, 119 S. Ct. 1692, 1701, 143 L. Ed. 2d 818 (1999). Justice Stevens dissented in part as it relates to another issue in the case.

8. evanescent evidence;
9. emergency;
10. miscellaneous other exceptions

Arrests

We have already discussed what an arrest is,[81] and have identified the requirements for obtaining an arrest warrant,[82] but have not discussed in what circumstances someone may be arrested without a warrant.

To arrest someone for a felony, no warrant is required if probable cause exists. If they are to be detained, law enforcement must bring the detainee before a neutral magistrate to determine if probable cause exists for the arrests. This meeting must occur within a reasonable length of time and no later than 48 hours after the arrest.

If probable cause is lacking, the arrest is invalidated, the detainee is released unless there is separate probable cause to keep them, and the prosecutor likely will not be able to use the evidence obtained from the arrest against the arrestee.[83]

If the crime committed is a misdemeanor, it must actually have been witnessed by the arresting officer for an arrest to be valid.

To enter someone's home to make an arrest, there must be probable cause and some exigent circumstance. In television shows and movies, the exigent circumstance is often pretextual. It's not unusual for two officers outside a door, after a knock and announce, to have an exchange similar to the following:

"Did you hear that?"

---

[81] *See* Arrest, page 98, above.
[82] *See* Arrest Warrant, page 99, above.
[83] *See* Exclusionary Rule in Chapter 12.

"Yeah, it sounds like someone calling for help."

Then they kick the door down.

Warrantless arrests in a home are presumed invalid, and the government has the burden of proving otherwise. Thus, if this were to occur in real life, a magistrate should invalidate a subsequent arrest when the officers failed to adequately explain the "exigent circumstances" that led them to enter the home and make an arrest without a warrant.

Searches Incident to an Arrest

When a person is legally arrested (i.e. with a warrant or under a valid warrant exception), police can search that person after the arrest. This exception is created to promote officer safety (to make sure any weapons or dangerous items on the arrestee are identified and confiscated) and to preserve evidence.

After an arrest, the officer can search not only the person arrested but can also search the area within the arrestee's wingspan—the area within his or her reach that might contain weapons or evidence. That includes the vehicle the arrestee is occupying for the period of time when the arrestee still has access to the vehicle (e.g., has not been handcuffed and placed in the back of the squad car), or even after the arrestee has been detained and separated from the vehicle if there is a reasonable belief by the police that evidence of the crime for which the person was arrested is in the vehicle.

This exception does not apply to data on a cell phone. This data presents no risk of danger to the officer, and there are greater privacy interests on the data on the phone such that to search it, there must either be a warrant or an independent exception that applies specifically to the phone's data.

The search-incident-to-arrest exception also does not apply where the person is not actually arrested. If it is merely an investigatory stop (being pulled over), a search is not authorized unless another exception exists.

The search incident to an arrest must occur contemporaneously with the arrest, meaning there is still a potential danger to police or risk of destruction of property.

Motor Vehicle Searches

For policy reasons, the law allows broader searches of motor vehicles than of homes or persons. First, there is a lesser expectation of privacy in a vehicle. Second, vehicles are mobile, so in the time it takes to get a warrant, it may no longer be available for a search. Additionally, motor vehicles, for obvious reasons, are not mentioned in the U.S. Constitution.

Thus, police can search an automobile if they have probable cause that the vehicle contains contraband (like drugs), fruits of a crime (like stolen money), instrumentalities of a crime (like a weapon used in a robbery), or evidence of a crime (like security guard routes or blueprints to a bank).

If probable cause exists to search the entire vehicle, police may search all areas in the vehicle, including the trunk and glove compartment, and any containers within the vehicle that might contain the item they're looking for. This may include containers, including purses, belonging not just to the driver, but also passengers.

If a vehicle has been impounded, it may be searched in its entirety without a warrant. This is allowed to protect the vehicle owners' property while it is in police possession by inventorying it and to protect the government should the owner later claim that the police stole, destroyed, or lost personal property contained in the vehicle when it was impounded.

The definition of a motor vehicle is broad and may include a motorhome (unless it is fixed to a certain spot and only used as a home, not as a vehicle).

Don't Look Under the Tarp

Ryan Collins got away with stealing a motorcycle because of the actions of some overeager police officers. This is another case that went all the way to the U.S. Supreme Court.

The motorcycle was black and orange and longer than most. And it was apparently a real joy to ride because what tipped the police off to Collins' presence was his reckless driving—at one point exceeding 140 mph.

Because of its superior acceleration and maneuverability, police were not able to apprehend Collins on the road, which they tried and failed to do twice.

Collins, in a move designed to reveal his own stupidity, posted photos on Facebook of the motorcycle parked in his girlfriend's driveway. The investigating officer, Officer Rhodes, tracked down her address and went to her house.

When Officer Rhodes arrived, he could not be certain whether the motorcycle was on the driveway—part of the home's curtilage—because all he could see was a motorcycle-shaped object covered by a tarp.

At that point, between the Facebook photo and the motorcycle-shaped tarp, he surely had enough to get a warrant. Instead, he entered the property and uncovered the tarp without a warrant and then subsequently arrested Collins.

Collins lost at trial and his appeal to the Virginia Supreme Court, where he argued the search was a violation of his Fourth Amendment rights.

The U.S. Supreme Court took the case and decided in its May 2018 decision that the automobile exception to the warrant requirement did not justify the invasion of the curtilage.

His Fourth Amendment rights had been violated.

Plain View

No warrant is required to seize an item if there is probable cause to believe that item is evidence, fruits, or instrumentalities of a crime, or contraband, as long as the police are legitimately on the premises when they discover it, and the item is in plain view (like, not under a tarp).

Consent

No warrant is necessary to conduct a search if a person appearing to have rights to use or occupy the property to be searched gives voluntary consent to the search.

Consent may be voluntary even if the person giving consent is not aware that they are free to refuse, but consent cannot be voluntary if the police lie about having a warrant.

If there is more than one occupant to an area to be searched, and the one against whom the search would be directed objects to the search, but the co-occupant consents, that consent is not valid. It is valid consent, though, if the co-occupant consents, and the person against whom the search is directed does not object, even if the failure to object is a result of that person's absence, including if their absence is due to a legitimate arrest after they did object to the search. I'm pretty sure this paragraph makes sense, even if only after reading it a few times. Phew.

Where consent is obtained, the search can only occur within the scope of consent, which includes all areas reasonably believed to be included in the consent.

Stop and Frisk

A stop and frisk is like a light version of a seizure and search, and thus has a lower standard of applicability.

A **stop** occurs when the police briefly detain someone for purposes of investigation, even though probable cause may not exist. It is only a stop if under the circumstances a reasonable person does not believe she is free to leave. It is like a mini seizure because it is brief and the inconvenience slight. It nonetheless falls under the ambit of the Fourth Amendment.

For a stop to be valid, the police must have reasonable suspicion supported by articulable facts. Reasonable suspicion falls on the spectrum somewhere between "vague suspicion" and "probable cause." If the police can competently articulate what caused them to effectuate the stop, it will probably be valid.

Stopping someone for being in a high crime area is not enough. Stopping someone who runs away at the sight of police is not enough. But together, these two factors are enough reasonable suspicion to stop someone.

A **frisk** is a police pat down to search for weapons. It is considered reasonable under the Fourth Amendment to promote officer safety, but only if the officer has a reasonable suspicion that the detainee is armed and dangerous.

Hot Pursuit

Where a felon flees the police, and the police are chasing the felon, this is considered a **hot pursuit**, which is an exigent circumstance allowing the police to seize and search the suspected felon without a warrant.

Evanescent Evidence

Evanescent means "quickly fading or disappearing." **Evanescent evidence** is evidence that, if not seized quickly, will disappear and have no value in the prosecution against the accused.

The evanescent evidence warrant exception applies in those cases where the evidence would disappear in the time it takes the police to obtain a warrant before seizing the evidence.

Drugs that can be flushed down a toilet, blood alcohol content that dissipates over time, and emails that can be easily deleted from anywhere, all may, under the right circumstances, qualify as evanescent evidence that can be seized without a warrant.

If the photograph that Marty McFly carried around with him in 1955 to remind him of the urgency in reacquainting his mother with his father were evidence of a crime—the one where he and his siblings were slowly being erased from existence, it too could meet the evanescent evidence exception.[84]

Emergency

Where there is some emergency that threatens safety or requires action to prevent injury, police can conduct warrantless searches. This exception would apply, for example, where police enter a home that is on fire or where there are objective signs that domestic abuse is occurring or has recently occurred. The emergency exception also applies to allow police to seize contaminated food to prevent it from being consumed.

Miscellaneous Other Exceptions

There are several other exceptions to the warrant requirement that are beyond the scope of this text, including roadblocks, border patrol, searches of the desks and files of government employees, government-required drug testing,

---

[84] *Back to the Future* © 1985.

searches of a parolee's home, searches of airline passengers, public school searches, and administrative inspections and searches.

CHAPTER TWELVE

The Exclusionary Rule

THE FOURTH AMENDMENT SETS THE limits on the government's ability to search and seize, but it does not outline what the consequences are for a constitutional violation. It took until 1914—123 years after the ratification of the Fourth Amendment, for the U.S. Supreme Court to create the exclusionary rule applicable in federal cases, which was the judicial branch's answer to consistent governmental violations of the Fourth Amendment. It wasn't until 1961, though, that the rule was made to apply uniformly in all states.[85]

The **exclusionary rule** prohibits the use of evidence against an accused in a criminal trial where that evidence was obtained in violation of the accused's Fourth Amendment

---

[85] *Mapp v. Ohio*, 367 U.S. 643, 81 S. Ct. 1684, 6 L. Ed. 2d 1081 (1961).

(unreasonable search and seizure), Fifth Amendment [86] (rights in criminal cases) or Sixth Amendment[87] (rights to a fair trial) rights.

The rule is meant to ensure governmental—mostly police—adherence to constitutional limitations by imposing consequences for the failure to do so. Because the police want to convict criminals, and their failure to follow the rules will make conviction more difficult (because the incriminating evidence can't be used), the exclusionary rule encourages police to follow the rules.

The rule not only applies to evidence illegally obtained, but also to most other evidence derived from the illegally obtained evidence. For example, where police illegally search a computer and find evidence of a bank account, and then they subpoena bank records, at trial, the prosecutor will not be able to use the computer records (because they were obtained illegally) or the bank records (because they were derived from information obtained illegally—without the illegally obtained information, the police never would have discovered it).

The rationale behind this rule, which is called **fruit of the poisonous tree** (or vine) is that if the tree is poisoned (illegally obtained evidence), then the fruit of that tree (evidence derived from illegally obtained evidence) is also tainted.

The exclusionary rule is not automatic, including the application of the fruit-of-the-poisonous-tree doctrine. The courts must apply a balancing test and weigh the culpability of the police against the probative value of the evidence. Thus, where the police are acting in good faith, but violate the rights of the accused in a serious crime, exclusion is more probable

---

[86] *See* Fifth Amendment, page 121.
[87] *See* Sixth Amendment, page 128.

than if the police are intentionally breaking the rules and the crime is less severe.

Additionally, there are four recognized exceptions to the exclusionary rule: independent source, inevitability, attenuation, and good faith.

Independent Source

Evidence obtained illegally, or tainted fruit from that evidence, is generally not admissible because we want the criminal system to have built-in incentives that encourage police to follow the rules. However, if the police find and seize evidence from a source independent of the illegal action, they will be able to use it against the accused.

Thus, if the police enter your home illegally when no one is home, and find contraband, they cannot use that contraband against you. Suppose they leave it there, knowing it has no evidentiary value. If the police collect enough evidence against you from other, independent sources, and get a valid warrant, they can then go into your home and gather the evidence. The independent source essentially cures the defect in the original search, which means the seized evidence can now be used against you.

Inevitability

Evidence that is not admissible because it was obtained in violation of the Constitution is nonetheless admissible if the prosecutor can prove at trial that the police would have inevitably discovered the evidence by lawful means.

Like other constitutional exceptions, this was born from a gruesome case.[88] Robert Williams escaped from a mental

---

[88] *Nix v. Williams*, 467 U.S. 431, 104 S. Ct. 2501, 81 L. Ed. 2d 377 (1984). This is another example of a dumb criminal shaping the law.

hospital and kidnapped and murdered a ten-year-old girl on Christmas Eve in 1968. He hid the girl's body and turned himself in at a police station in a neighboring county on the condition that the police ask him no questions during the trip to his home county. The police agreed.

During the drive back, the police asked him if he would disclose the location of the girl's body so they could find it before it started snowing. He agreed and led them to it.

At trial, his attorneys argued that the incriminating statements, his knowledge of the body's whereabouts, and the body itself, could not be used against him because they were obtained from an illegal confession.

On appeal, the U.S. Supreme Court held that the incriminating statements and his knowledge of the body's whereabouts were not admissible against him as they were obtained illegally. And although the location of the body was discovered from the illegally obtained confession, because the police's discovery of the body was inevitable (the snow was soon going to melt and reveal the body), it could be used against Williams, who was eventually convicted.[89]

Attenuation

Where the illegal police act that made the evidence inadmissible and the corresponding discovery of new evidence are far enough removed from each other that the causal chain is sufficiently attenuated, and the taint (mostly) removed, the new evidence can still be used.

This often involves an act of free will by the defendant, like where the defendant's confession is coerced illegally (which can't be used against her), but then she returns later

---

[89] Notably, the crime happened in 1968, and after two trials and two appeals, the U.S. Supreme Court finally upheld the conviction in 1984—sixteen years later.

to the police of her own free will and confesses again. The first confession cannot be used. The second can.

Good Faith

Where the police execute a search warrant, with a good-faith belief that the warrant is valid but the warrant turns out to be invalid, the evidence can still be used at trial. This exception turns on whether the mistake was made by the police seeking and executing the warrant or by the judge issuing it.

Because the exclusionary rule is designed to promote proper police conduct, not judicial conduct, if the police have done everything right, and the magistrate makes a mistake in issuing a warrant, the evidence will still be admissible. On the other hand, if the police mislead the magistrate to get the magistrate to issue a warrant where she wouldn't otherwise, or if the police realize the warrant should not have been issued, but execute on it anyway, the good-faith exception will not apply, and the evidence will be inadmissible.

LIMITATIONS

The exclusionary rule is not universal in its application. Besides certain factually specific exceptions outlined above, there are several settings where the rule does not apply at all.

The exclusionary rule does not apply in grand jury proceedings,[90] for example.

It also does not apply in civil cases. Thus, evidence that is illegally obtained, and which is never admitted into evidence in the criminal trial, can still be used in a civil trial against the accused.

---

[90] *See* Grand Juries, page 135.

Because the exclusionary rule was created to promote constitutional protections, it does not apply where state law or administrative regulations are violated. Or where the constitutional rights of someone other than the accused are violated.

The rule also does not apply to evidence unlawfully obtained by a private person, since the rule is designed to encourage proper conduct by government officials, not private citizens.

Finally, the scope of the exclusionary rule does not extend to protecting foreigners who are outside the U.S. borders or in proceedings to determine whether parole will be revoked.

CHAPTER THIRTEEN

Interrogation and Confessions

OF ALL THE PRIMETIME PORTRAYALS of criminal investigation and prosecution, perhaps none is so dramatic or oft portrayed as the interrogation. In crime dramas, there may be several interrogations in the same episode: every interrogated suspect has an alibi, then the police find out that one of the suspects lied, they confront that suspect about the lie, but there is a good explanation for that, too, and in the end, the person who actually commits the crime is the one the viewer least expected, whose guilt is revealed by a spontaneous confession elicited in less than a minute by a very clever line of questioning by the police. Sound familiar?

Almost every one of these crime-drama interrogations has one of three attributes: (1) the suspect does not ask for an attorney, (2) the request for an attorney is ignored, or (3) there is an attorney present, but that attorney is portrayed as incompetent and annoying.

This anything-it-takes-to-get-a-confession approach makes for good television, but in real life, the Constitution protects a suspect throughout the interrogation, and there are consequences for a police officer's violation of these rights, including the suppression of the confession.

A suspect's rights during interrogation find their origin, in addition to the Fourth Amendment, which has already been covered, in the Fifth, Sixth, and Fourteenth Amendments.

FIFTH AMENDMENT AND MIRANDA

In a literal sense, this part of the **Fifth Amendment** means a

> *Amendment 5: No person ... shall be compelled in any criminal case to be a witness against himself....*

criminal defendant does not have to testify—and cannot be compelled to testify—at his own criminal trial. But this constitutional right is broader than that. It also means that all persons (whether or not a defendant) have a right not to make incriminating statements (statements that implicate them in the commission of a crime), and they cannot be compelled to give self-incriminating testimony.

Fifth Amendment rights are waivable, though. Suspects in a criminal investigation (or anyone, for that matter), if they choose, may confess to crimes they've committed (and a confession is incriminating by definition).

But a long history of police abuse of constitutional rights has created a need for courts to question the validity of any confession. To be valid, and therefore admissible, a confession must represent a voluntary, knowing, and intelligent waiver of constitutional rights. Enter Ernesto Miranda.

Ernesto Arturo Miranda, an Arizona native who had a significant criminal history as a minor, became a person of interest in a kidnapping, rape, and armed robbery case in 1963 when he was 22 years old.

Lois Ann Jameson, then 18, was returning home from a late shift when an unknown assailant approached her with a knife and forced her into his car, which he used to take her out to the desert, where he raped her and took her money. He then drove her back towards the city and dropped her off a few blocks from her home.[91]

Miranda's girlfriend's car

Jameson had the presence of mind to note her attacker's attributes, the attributes of the green car he was driving, and make a mental note of the license plate, all of which she shared with the police and her brother.

Her brother spotted the vehicle and informed the police, who investigated. The car belonged to Miranda's live-in girlfriend, and at the request of police, Miranda voluntarily went down to the station.

He was placed in a lineup with three other men, and Jameson identified him, but admitted she was not completely sure it was him.

The police told Miranda he'd been positively identified, took him to interrogation room 2, and began what would be a two-hour interrogation. The police did not inform Miranda that he had constitutional rights he could assert.

---

[91] In some twisted way, he probably thought he was being chivalrous to give her a ride back from the desert.

Ernesto Miranda is no. 1 in this lineup.

At the end of the interrogation, Miranda confessed not only to the rape, but also to eight other unsolved crimes. He was taken to meet Jameson, where he said "That's the girl," and she confirmed that his voice matched the voice of her assailant.

The police gave him a standard confession form, which at the top stated, "I ... do hereby swear that I make this statement voluntarily and of my own free will, with no threats, coercion, or promises of immunity, and with full knowledge of my legal rights, understanding any statement I make may be used against me." He filled out the confession form with details of the rape.

At Miranda's trial, his attorney objected to the admission

of the confession, but the judge overruled the objection and allowed the jury to consider it. The jury convicted Miranda of rape and kidnapping. He was sentenced to 20–30 years.

The Arizona Supreme Court upheld the conviction on appeal. While imprisoned, Miranda, through a new set of attorneys, appealed to the U.S. Supreme Court alleging that his Fifth and Sixth Amendment rights were violated.

In a well-reasoned but divided decision, the High Court took the opportunity to clarify the extent to which a suspect must be informed of their rights, which have since become known as **Miranda Rights**: "Prior to any questioning, the person must be warned that he has a right to remain silent, that any statement he does make may be used as evidence against him, and that he has a right to the presence of an attorney, either retained or appointed."[92]

The Supreme Court concluded that Miranda's constitutional rights had been violated. It overturned the conviction.

The state of Arizona tried Miranda again, this time without introducing the confession into evidence. The prosecution was still able to secure a conviction, and Miranda was again sentenced to 20–30 years in prison.[93]

This single decision, which even after 52 years is still controlling law, permanently changed law enforcement procedure. Now it is common practice for the police to inform a suspect of their rights by "**Mirandizing**" them by reading or reciting to them their Miranda Rights (most of which you can probably recite):

---

[92] *Miranda v. Arizona*, 384 U.S. 436, 444, 86 S. Ct. 1602, 1612, 16 L. Ed. 2d 694 (1966).
[93] Miranda served nine years of his sentence and was paroled in 1972, at which time his name had become a verb ("Mirandize"), and he was famous. Capitalizing on his infamy, he began selling autographed Miranda warning cards for $1.50 each. He died in 1976 after getting stabbed in a bar fight. He was 34.

You have the right to remain silent. Anything you say can and will be used against you in a court of law. You have the right to an attorney. If you cannot afford an attorney, one will be provided for you.[94] *Do you understand the rights I have just read to you? With these rights in mind, do you wish to speak to me?*

This duty to Mirandize a suspect kicks in once the suspect is in custody, meaning they are no longer free to leave. It doesn't have to be in an interrogation room. It can be out in public, in their vehicle, or while they are at home.

This duty only applies to governmental conduct. A confession to a police informant is valid even without Miranda warnings. Additionally, self-incriminating testimony by a witness at a grand jury proceeding can be used against a witness who has not been Mirandized.

Only after the detainee has been informed of their rights can a waiver of those rights be a knowing waiver. At that point in an interrogation—after the suspects know they don't have to answer questions, and they can ask for an attorney—are their confessions valid as incriminating evidence against them (as long as the confession is also voluntary, discussed below).

The right to refuse to talk to police (remaining silent), can be invoked even after it has been waived. In fact, if at any point before or during the interrogation, the suspect can invoke that right. Silence alone is not an invocation, but where the suspect says he no longer wants to answer questions (or some other indication that he is done talking or wishes to remain silent), all efforts to solicit a confession must stop. The interrogation is over.

---

[94] It's usually at about this point in the warnings on television cop dramas that the scene cuts away.

The Fifth Amendment Right to Counsel

Although the Fifth Amendment makes no explicit mention of a right to counsel, the *Miranda* court inferred one, and ever since *Miranda*, there has been a Fifth Amendment right to counsel.

The Fifth Amendment right to counsel only applies during a custodial interrogation. Thus, the analysis of this right is similar to the right to remain silent. Any confession volunteered by a suspect in custody is inadmissible unless they have been read their rights, including the right to remain silent and the right to have an attorney.

If the detainee at any point asks to speak to an attorney, police can ask no further questions until either (1) the suspect has been afforded an attorney, or (2) the suspect re-initiates the questioning.

FOURTEENTH AMENDMENT

Amendment 14 § 1: ... nor shall any state deprive any person of life, liberty, or property, without due process of law ....

In our country's early years and before (most notably in the Middle Ages), confessions were coerced by means of physical torture. Early interrogators in the United States would place a wooden board on top of the suspect and then continually add heavy stones, slowly crushing the suspect under the increasing weight. The suspect was told they merely had to confess to the crime, and the increasingly heavy weights would be removed. As a method for eliciting confessions, it was very effective—having the life crushed out of the suspects was a great motivator and resulted in confessions for anything, including crimes they had no part of.

Historically, several systems of criminal justice have been based on confessions as a means of proving guilt, rather than putting the burden on the government to acquire independent evidence to prove guilt without a confession. Think back to the Spanish Inquisition—a centuries-long crusade to root out heretics that resulted in about 150,000 prosecutions, many of which resulted from confessions extracted by torture.

The Due Process Clause of the **Fourteenth Amendment** has been interpreted to include a requirement that a suspect's confessions be voluntary, which disallows, among other things, confessions coerced by means of crushing the human body and other physical torture. The rule on voluntariness is simple: if confessions are not voluntary, they are not admissible.

With the rule now established, the question becomes, what constitutes a voluntary confession? Like almost everything else in the law, there is no easy answer.

First, it is up to the prosecution to prove that the confession was voluntary. Courts will presume the confession is not valid until the government convinces them otherwise.

To determine voluntariness, courts will consider the confessor, including their age, education level, mental and physical condition, and whether they speak English. Courts will also examine the circumstances surrounding the interrogation, including where it occurred, how long it lasted, and the methods used by police.

For example, any promises made by the police will be closely scrutinized. And although the police don't have to tell the truth in an interrogation, a lie cannot be used to circumvent constitutional protections. For example, telling the suspect that their buddy in the next room has already confessed is allowed. Telling the suspect that if they confess they will be able to go home is not. Telling the suspect that they have a video of the crime being committed is fine. Telling the suspect that if they do not confess, the police cannot protect their family is not.

A confession elicited by an undercover cop who tells the suspect that they are a priest, attorney, or doctor (or anyone with whom a conversation would be privileged) is neither voluntary nor admissible. On the other hand, if the suspect spills the beans to his cellmate who happens to be a police informant, those statements are fair game.

SIXTH AMENDMENT

Amendment 6: In all criminal prosecutions, the accused shall enjoy the right ... to have the Assistance of Counsel for his defence.

The **Sixth Amendment** right to counsel seems pretty straightforward—criminal defendants have a right to a defense attorney—but it wasn't always applied that way.

In fact, it wasn't until 1963 that the U.S. Supreme Court declared the rule, and only after a man was convicted after he represented himself at trial.

In the wee hours of June 3, 1961, an unknown person broke into a pool hall in Panama City, Florida. This person destroyed a cigarette machine [95] and record player [96] and emptied the cash register.

A witness reported having seen one Clarence Earl Gideon exiting the pool hall at 5:30 a.m., which tip resulted in his arrest.

Gideon appeared at court on his own behalf and asked the judge to appoint an attorney to represent him since he couldn't afford to hire an attorney. The judge declined, citing Florida law, which only required an appointment of counsel in cases where the death penalty was sought.

Gideon defended himself at trial, steadfastly maintaining his innocence. Notwithstanding his efforts, a jury convicted him, and he was sentenced to a five-year prison term.

Once incarcerated, Gideon made use of the prison library, picked up a pencil, and wrote his own appeal. He challenged the Florida judge's decision to deny him counsel, arguing that it was a violation of his Sixth Amendment right.

The U.S. Supreme Court, with some irony, appointed an attorney to argue an appeal on his behalf.

The Court ultimately sided with Gideon, agreeing that the Sixth Amendment did not distinguish cases based on the punishment sought, and a defendant charged with any crime, who could not afford an attorney, must have one appointed for him.

The U.S. Supreme Court's decision resulted in a new trial for Gideon, this time with an attorney. The jury acquitted him of all charges after one hour of deliberation.

---

[95] Cigarette machines are cigarette vending machines. They are only allowed in facilities that restrict entrance to people 18 years old or older.

[96] Record machines are ancient music-playing devices where a vinyl disc was placed on a rotating turntable where analog signals were sent via a needle to an amplifier and speakers. Music purists believe the best way to hear music that isn't live is still from a record.

As a result of this decision, not only were approximately 2,000 people released from prisons in Florida, but ever since, every indigent defendant in any state has had the right to be represented by an attorney.[97]

This Sixth Amendment right to counsel applies during any critical stages of a criminal prosecution, which includes the formal charge, preliminary hearing, information, indictment, arraignment, plea negotiations, entering a guilty plea, and any line-up or interrogation that occurs after an indictment. It does not necessarily include an arrest, though *Miranda* broadened the right to an attorney, based on the Fifth Amendment, to include any time after detention when the detainee asks for one—and of course, they have to be made aware that the option is available.

---

[97] The case is *Gideon v. Wainwright*, 372 U.S. 335, 83 S. Ct. 792, 9 L. Ed. 2d 799 (1963).

CHAPTER FOURTEEN

Pretrial Procedures

IT IS USUALLY LAW ENFORCEMENT that investigates the crime, makes the arrests and brings the accused to the police station for booking, questioning, and processing. The police keep records of their investigations and prepare written reports summarizing their findings and conclusions (it's that paperwork police on television are always complaining about but never actually doing on camera).

The results of this investigation are presented to a prosecutor. The **prosecutor** is an attorney who works for the government and whose job it is to assess the evidence against a given suspect, exercise discretion in deciding whether to pursue charges, and if so, see the case through to the end, which means either a dismissal, a plea, a conviction, or an acquittal.[98] The prosecutor will be one of many attorneys working for the District Attorney for the state (also known by other similar names, like State's Attorney or People's Attorney).

If charges are going to be brought, the prosecutor signs and files the criminal complaint against the accused, at which point the suspect graduates to defendant status.

The complaint is either accompanied by a summons (requesting the defendant's presence at court) or an arrest warrant (which police use to detain the defendant and compel attendance), depending on how likely the prosecutor thinks it is that the defendant will appear willingly if summoned.

The defendant must then appear at the initial appearance. The **initial appearance** is a hearing before a judge where defendants are made aware of their constitutional rights, including the right to counsel, and they are informed of the charges being brought against them. The court also ensures that the defendant has not been illegally detained by the police.

The initial appearance must occur shortly after an arrest. The time limit varies from jurisdiction to jurisdiction, but in all jurisdictions, under normal circumstances, no more than 48 hours can pass between the arrest and the initial appearance.

Also often determined at the initial appearance is the issue of bail. **Bail** is the security pledged, either monetary or

---

[98] This process is portrayed in those special two-hour *Law & Order* crossover episodes, where for the first hour, the Special Victims Unit will investigate the crime that eventually ends in an arrest, and then they hand the suspect over to the prosecutor for the second hour, where the prosecutor has to overcome several legal obstacles to secure the conviction.

promissory, to ensure the defendant's attendance at later hearings, in exchange for the defendant's release from jail pending the disposition of the case.

There are two main issues a court must consider to decide whether the defendant will be released from custody: (1) securing the defendant's attendance at future hearings and the trial, and (2) protecting the community from danger should the defendant be released.

Thus, at a bail hearing, the judge will hear arguments regarding whether the defendant is a flight risk, including the seriousness of the crimes being charged, the strength of the evidence of the crime, the defendant's ability to post bail, the defendant's ties to the community, means of departure, etc. It will also consider whether the defendant poses a danger to the public.

Based on an analysis of these factors, the judge will do one of the following:

- Release on **personal recognizance**. Where the defendant is not a flight risk and poses no danger to the community, the court may let the defendant go on the promise of future attendance alone.
- Release upon **posting of bail**. Different jurisdictions handle this differently, but generally the posting of bail requires a deposit of some sum of money with the court (or sometimes just the promise to be indebted to the court if the defendant does not appear), which can be paid by the defendant or someone other than the defendant, including friends, family, or bail and bonding companies.[99] That money is returned if the

---

[99] Because a bonding company has a financial stake in whether the defendant appears, they will do what they can to ensure the defendant shows up when required, including

defendant continues to appear through the end of the trial, but if at any point the defendant does not appear at a required hearing or trial, the judge can order the bail forfeited. The greater the flight risk or danger to society, the higher the bond will be set.
- **Deny bail.** If the court determines that the flight risk is so great that no amount of money will be enough to induce the defendant to appear, or the danger to society is too grave, it can deny bail altogether and require the defendant to be detained until acquittal or the end of any imposed sentence after conviction.

Defendants who can afford to hire an attorney and choose to do so likely will have hired an attorney before they have to appear at the initial appearance.

The governing body charging the defendants, if it is proven that the defendant cannot afford an attorney (which is determined by specific financial criteria), provides an attorney, usually at or before the initial appearance. In jurisdictions with a public defender's office, the case will be assigned to one of the attorneys working for the public defender. In less populated jurisdictions that do not have a public defender, private attorneys are court appointed to represent criminal defendants. The government pays these attorneys a fee for their services.

PRELIMINARY HEARING

Recall that police must have probable cause to arrest

---

hiring bounty hunters to track down bail jumpers and haul them back to court to get their bail deposit back.

someone. Shortly after the arrest, the defendant is entitled to a probable cause hearing where a judge reviews the evidence to make sure there is probable cause to believe that the defendant committed the crime they are accused of.

This probable cause hearing is called a **preliminary hearing.** If the arrest was made pursuant to an arrest warrant, then probable cause has already been determined, and a preliminary hearing is unnecessary. The same is true if the arrest was made as a result of a grand jury's indictment. Additionally, if the defendant has been released on her own recognizance, there is no detention, so preliminary hearing is unnecessary.

At the preliminary hearing, the judge reviews the evidence against the defendant, which usually consists of the testimony of the police, but could also include testimony from other witnesses or other tangible evidence. If the judge determines that there is probable cause for the arrest and detention (i.e. that the crime was committed by this defendant), then the court transfers the case to the next court, called **binding over**. If there is no probable cause, the defendant is released and the charges dismissed.

The preliminary hearing and the initial appearance may occur at the same time.

INDICTMENTS AND GRAND JURIES

Amendment 5: No person shall be held to answer for a capital, or otherwise infamous crime, unless on a presentment or indictment of a Grand Jury ....

Where the Fifth Amendment says "otherwise infamous crime," we interpret that to mean a felony. Interestingly, this provision of the Bill of Rights, unlike many others, applies only to the federal government, not to the states. Regardless, most states, but not all, do use grand juries[100] to some extent.

A **grand jury** is a group of U.S. citizens convened to hear evidence and decide whether there is enough evidence to allow a trial against the accused to go forward.

In many jurisdictions, including the federal system, a case must either go through a grand jury, a preliminary hearing, or both. In the federal system, for example, a grand jury must establish that there is probable cause to go to trial for felonies, and a judge at a preliminary hearing must determine that there is probable cause for misdemeanors.

For felony cases, the prosecutor drafts a charging document called an indictment. The **indictment** is a document with a short and plain statement outlining the facts comprising the offenses the defendant is being charged with, with each offense set out as a separate count. In cases where a grand jury is not necessary, or where the defendant waives

---

[100] A grand jury, which consists of 12 to 33 members (between 16 and 23 in federal cases), makes determinations of probable cause to prosecute. A petit jury renders a verdict in a civil or criminal case. When you think of a typical jury, you're thinking of a petit jury.

the grand jury requirement, the charging document is called an **information**, which is obtained either from the presentation of evidence at a preliminary hearing or by consent of the defendant.

Investigative grand juries (also called special grand juries) can perform their own investigations into crimes and bring indictments on their own initiative.

The prosecutor takes the indictment to the grand jury and then presents the evidence against the accused. The grand jury proceedings are secret. The defendant does not know the grand jury has been convened or is considering a case against the defendant. Nor does the defendant have access to the proceedings. The defendant cannot present evidence, the prosecution is not required to present any exculpatory evidence, and grand juries can consider information that would be excluded at a trial.

Witnesses at a grand jury proceeding do not have to have their *Miranda* rights read to them, do not have the right to an attorney, and if called to testify, do not need to be given notice that they are the person of interest for the criminal case.

Grand juries can issue subpoenas on their own, which don't require probable cause, and which are not subject to the same scrutiny as other subpoenas.

If, after considering the evidence, the grand jury believes there is probable cause to proceed to trial against the accused, votes "**true bill**," and signs the indictment, making it official. If the relatively low burden of probable cause is not met, the jury votes "**no bill**," and the case against the defendant cannot continue.

As a practical matter, because prosecutors have the discretion to decide which cases to pursue, and because the probable cause standard before a grand jury is a relatively low bar, grand juries almost always issue the indictments. Prosecutors typically don't bring cases before the grand jury that they don't have the evidence to support.

In some cases, even though the prosecutors do not believe they have enough evidence, they may bring the case before a grand jury anyway, including where there is great public interest.

In early 2018, eight officers in Savannah, Georgia fatally shot Ricky Boyd, a 20-year-old black man. There was some question after the shooting about whether Boyd had a weapon in hand when police shot him, with initial reports claiming Boyd shot at them first and then confronted the officers with the gun. It was later determined that Boyd was brandishing a BB gun when the officers approached him.

The case was brought before a grand jury, which reviewed statements made by Boyd's family to law enforcement, watched body cam footage from two of the police, and heard various witness testimony. The grand jury did not issue the indictment, providing a report with its conclusion: "It is the opinion of the grand jury that the shooting of Ricky Boyd was justified as being reasonably necessary to prevent Ricky Boyd's perceived imminent use of unlawful deadly force against law enforcement officers nearby."

Recall the famous D.B. Cooper case discussed above?[101] In 1976, five years after the famous hijacking, prosecutors became concerned that the statute of limitations would run before D.B. Cooper could be apprehended and charged. In a creative move, Portland prosecutors convened a grand jury, presented evidence, and were rewarded with an indictment in absentia against "John Doe, aka Dan Cooper." The charges?

---

[101] *See* page 52.

Air piracy and robbery affecting interstate commerce in violation of the Hobbs Act. Because the case has been formally initiated, if at any point in the future his identity is ascertained, they can pick up the case where they left off over 40 years ago with no concerns about expired statutes of limitation.

ARRAIGNMENT

Where an indictment or information is obtained, the next procedural step is the arraignment. The **arraignment** is a hearing where the judge informs the defendant of the charges against him, informs the defendant of his rights, and gives the defendant an opportunity to enter a plea in response to the charges. The defendant can enter one of three pleas:

- **Not guilty.** If a defendant pleads not guilty, a trial date is set, and the government must prove its case against the defendant.
- **Guilty.** If the defendant pleads guilty, the court asks a series of questions of the defendant designed to satisfy the court that the defendant understands the implications of the plea, is entering the plea knowingly and voluntarily, and is aware of the constitutional rights being forfeited with a guilty plea, including the right to a jury trial, to confront witnesses, and to remain silent. The judge must also be satisfied that there is some factual basis to the charges. The judge does not have to accept the guilty plea. If the guilty plea is accepted, the defendant proceeds to sentencing.
- **No Contest** (*nolo contendere*). A no contest plea is the

same as a guilty plea for all purposes in the criminal case. It has the effect of the defendant accepting the consequences of a guilty plea without actually admitting guilt. This prevents the criminal consequences of the defendant's behavior from being used against the defendant in a civil trial for the same acts.

The reality is, most cases end in a negotiated plea bargain between the prosecution and the defense. A **plea bargain** is a deal where the prosecutor offers to charge the defendant with a lesser offense than those brought against the defendant in exchange for the defendant's plea of guilty to the lesser charge. There are so many cases that a prosecutor could not possibly take them all to trial (nor could the public defender), resulting in economics-driven plea bargains (scarcity of time and resources).

SPEEDY TRIAL

Amendment 6: ...the accused shall enjoy the right to a speedy and public trial....

Another important right that the Constitution endows on criminal defendants is the right to a speedy trial, if they so choose. This right vests once the defendant becomes accused in some way, such as when the indictment is issued.

But a defendant may not want to assert her constitutional right to a speedy trial for any of a number of reasons. The passage of time, with the accompanying fading of memories, for example, may have a more negative impact on the prosecutor—who has the burden of proof—than on the defendant.

A defendant gets to choose whether to assert the right to a speedy trial, which is done by serving a speedy trial demand on the prosecution. When that happens, the trial is usually set very quickly.

Although there is no black-and-white rule about how long is too long to wait for a trial in those cases where a speedy trial demand is not served, the U.S. Supreme Court has established a set of factors to be considered when determining whether the right to a speedy trial has been violated:

(1) Length of the delay;
(2) Reason for the delay (did the defendant cause it?);
(3) Whether the defendant asserted the right to a speedy trial; and
(4) Whether the delay resulted in prejudice (undue or unfair harm to defendant's case).

There is but one remedy for the violation of a defendant's right to a speedy trial: dismissal of the case.

DISCOVERY

In civil cases, discovery is one of the most important and time-consuming stages of litigation. Although discovery processes vary widely from jurisdiction to jurisdiction, historically, discovery has played a very different role in criminal cases than it has in the civil arena.

Criminal discovery has traditionally been a burden borne by the prosecutor alone. The prosecutor has the duty to turn over a list of the witnesses that would testify against the defendant, documents and other tangible objects, statements made by the defendant and any co-defendants, defendant's criminal record, and reports from examinations and tests (ballistics, DNA, etc.).

The prosecutor also must turn over any exculpatory evidence, or evidence that would help the defendant in her case and might exonerate her. This is called *Brady* material, named after John Leo Brady's landmark U.S. Supreme Court case, *Brady v. Maryland*.[102]

John Leo Brady was born in Dunkirk, Maryland in 1932. Unrelated to his case, but nonetheless interesting, he earned himself the nickname, "old stinkears" due to a constant seeping pus coming out of his ears for the first two decades of his life.

As a young adult fresh out of the Air Force, Brady found love and got her pregnant. Pledging to take care of his unborn child, Brady wrote a check for $35,000 to his girl but told her not to cash it for two weeks. In the next fourteen days, his plan was to rob a bank to back the check and support his growing family.

Brady and his partner in crime, Donald Boblit—his girlfriend's brother—needed a getaway car for the robbery. Brady grabbed his pistol and Boblit his shotgun, then laid a log across the road outside the home of William Brooks, who owned a vehicle and was at work.

Brooks stopped his vehicle as he approached his home and got out of his vehicle to move the log. That's when Boblit knocked him over the head with a shotgun. The two loaded him up in his car and took him up to an abandoned field, where Boblit strangled Brooks to death.

In the time leading up to the trial, Boblit confessed five different times. The first four times, he testified that Brady killed Brooks. The last time, he admitted that he had done the deed.

---

[102] 373 U.S. 83, 83 S. Ct. 1194, 10 L. Ed. 215 (1963).

Brady asked for a jury trial, but Boblit was content with a bench trial. Brady's was first, and his attorney asked for Boblit's statements from the prosecutor. He was given the first four, but the prosecutor withheld the fifth. At trial, Brady confessed to his role in the crime, but insisted that Boblit had done the actual killing. Unconvinced, the jury convicted him, and he was sentenced to death.

At Boblit's trial, the prosecutor attempted to introduce the fifth confession, but it was deemed inadmissible because it was not signed (which was required under Maryland law). Boblit, too, was convicted and sentenced to death.

While on death row, a Jesuit priest sought and found a new attorney to represent Brady. This attorney could not at first find a basis for an appeal, but then reviewed the transcript for Boblit's trial. That's when he found out that there was a fifth confession that if true, could have made things better for Brady. He wanted a new trial.

Both the appeals court and the U.S. Supreme Court ruled that Brady's Due Process rights were violated, and he was denied a fair trial. However, because (1) the confession would have been inadmissible (it wasn't signed, remember), and (2) he would still have been convicted under the felony-murder rule, the conviction was affirmed. However, the U.S. Supreme Court created a rule that the prosecution must, without being asked, turn over all "exculpatory evidence" to the defendant. And it is this rule that *Brady* is known for, and which has changed a prosecutor's role in all criminal cases since.

Because the Maryland courts did not yet have a process for a sentencing-only non-trial, the case languished. And not wanting to risk being sentenced again to death, Brady, who had been transferred from death row to general population in prison, remained content, for the time being, an unsentenced prisoner.

After 18 years, Brady's attorneys figured enough time had passed that it would be difficult for the prosecutors to make a strong case, so they asked for a speedy punishment proceeding. That prompted the governor's involvement, clemency, and release. Brady became a truck driver, married, and started a family. He had a clean record going forward, having made his mark in the world helping defendants all over the country.

CHAPTER FIFTEEN

Trial

TRIAL IS THE CULMINATION OF the criminal judicial process. The arrest, charges, discovery, plea deals, etc., all are shaped and driven by what proof is required at trial and the attorneys' estimation of what will happen at trial. Yet, according to the United States Courts' website, "[m]ore than 90 percent of defendants plead guilty rather than go to trial."[103] Of those that do go to trial, at least according to one source, "about 90 percent end in a guilty verdict."[104]

That means approximately 99% of all defendants charged either admit guilt or have guilt proven against them. Some see this as evidence that the system is rigged against

---

[103] United States Courts, "Criminal Cases," available at http://www.uscourts.gov/about-federal-courts/types-cases/criminal-cases (last accessed July 31, 2018).

[104] Jason Trahan, "'Not Guilty' Remains a Rarity in Federal and State Courts," *Dallas Morning News*, January 29, 2012, available at https://bit.ly/2HOts5Y (last accessed July 31, 2018).

defendants. Others see it as an indication that the prosecutors are exercising their discretion wisely and are only bringing charges when they can actually prove the defendant's guilt.

Regardless of which side is right (and there is probably truth to both positions), there are many constitutional guarantees built into the system to try and ensure that defendants get a fair shake. We have already discussed a number of those, including the right to remain silent and the right to an attorney. In this chapter we will discuss the right of a defendant to a public trial, to a jury trial, to confront witnesses, and rights regarding the burden of proof.

RIGHT TO PUBLIC TRIAL

Amendment 1: Congress shall make no law ... abridging the freedom ... of the press ....

Amendment 6: ...the accused shall enjoy the right to a speedy and public trial....

Amendment 14 § 1: ... nor shall any state deprive any person of life, liberty, or property, without due process of law ....

We don't normally think of publicity of judicial proceedings as a right, but it was something our Founding Fathers built into the Constitution. Interestingly, it is fashioned both as a right of the accused and of the public, though the press.

The U.S. Supreme Court has described the importance of the right to access the trial: "The right to attend criminal trials is implicit in the guarantees of the First Amendment; without the freedom to attend such trials, which people have exercised for centuries, important aspects of freedom of speech and of the press could be eviscerated."[105]

However, sometimes these competing rights can be at odds with one another, and the press' access to the trial can interfere with the defendant's constitutional rights. Nothing illustrates this principle as well as the highly publicized trial of Billie Sol Estes.

Estes v. State of Texas

Estes was a Texas businessman who, at age 28, was honored to receive the "One of America's 10 Outstanding Young Men of 1952" from the United States Junior Chamber of Commerce.

Estes, a budding entrepreneur, cooked up a brilliant fraud scheme. He would sell anhydrous ammonia[106] tanks to farmers, convincing them to take out a mortgage on the tanks, sight unseen. The farmers, familiar with these tanks, purchased the mortgages on credit because Estes told them he would lease the tanks from them for the same price as the mortgage payment plus a convenience fee. Thus, from the farmers' perspective, they didn't have to come out of pocket, and would even make some money, so they were basically able to cash in on their good credit.

The problem was, these tanks never existed.

Estes then sought loans from banks outside of Texas using the fraudulent mortgage holdings as collateral.

After a few years of running this and other similarly

---

[105] *Richmond Newspapers, Inc. v. Virginia*, 448 U.S. 555, 556–57, 100 S. Ct. 2814, 2817, 65 L. Ed. 2d 973 (1980).
[106] Anhydrous ammonia is used as an agricultural fertilizer.

clever scams, he bribed four officials at the Department of Agriculture, which led John F. Kennedy to launch an investigation into Estes' activities. That investigation revealed several business dealings, including some with Lyndon B. Johnson, who would later become president.

Estes was tried on several counts of fraud, among other things, and his trial was highly anticipated by the public.

> *The two-day pretrial hearing was "carried live by both radio and television, and news photography was permitted throughout. The videotapes of these hearings clearly illustrate that the picture presented was not one of that judicial serenity and calm to which petitioner was entitled. Indeed, at least 12 cameramen were engaged in the courtroom throughout the hearing taking motion and still pictures and televising the proceedings. Cables and wires were snaked across the courtroom floor, three microphones were on the judge's bench and others were beamed at the jury box and the counsel table. It is conceded that the activities of the television crews and news photographers led to considerable disruption of the hearings."*[107]

Because of concerns caused by the interference of the press at the pre-trial hearing, the trial was transferred to a venue several hundred miles away, and the press was only given limited access to the actual trial.

The jury that was empaneled for trial included four jurors that had been exposed to all or part of the broadcasts of the pretrial proceedings.

The jury convicted Estes, and he received a 24-year prison sentence. On appeal, Estes argued that he was deprived of a fair trial due to all the publicity.

---

[107] *Estes v. State of Tex.*, 381 U.S. 532, 536, 85 S. Ct. 1628, 1629, 14 L. Ed. 2d 543 (1965).

The Supreme Court explained the purpose of the right to a public trial: "The purpose of the requirement of a public trial was to guarantee that the accused would be fairly dealt with and not unjustly condemned. History had proven that secret tribunals were effective instruments of oppression."

Recognizing that the rights of the press are also important, and conflicted in this case with the rights of the accused, the Court announced that "the primary concern of all must be the proper administration of justice; that the life or liberty of any individual in this land should not be put in jeopardy because of actions of any news media; and that the due process requirements in both the Fifth and Fourteenth Amendments and the provisions of the Sixth Amendment require a procedure that will assure a fair trial."

The Court cited several examples of how the proceedings as they were conducted undermined Estes' due process rights and denied him a fair trial, including the impact of televising a trial on the jury, the "nerve-center of the fact-finding process," increasing the potential for prejudice; the impact of television a trial on the witnesses, who, knowing the whole country is watching, may be demoralized and frightened or cocky and given to overstatement; the presence of a television crew and equipment on the difficult job of the judge; and finally, the impact the television equipment has on the defendant, which may resemble harassment and impose on his dignity.

The Court reversed the judgment below by a narrow, 5-4 margin. However, in a prescient conclusion to the opinion, the Court recognized that as technology improves, the "telecasting" in courtrooms may be less intrusive and interfering with justice in a trial. "But we are not dealing here with future developments in the field of electronics. Our judgment cannot be rested on the hypothesis of tomorrow but must take the facts as they are presented today."

Estes would later strike again and be convicted of new fraud charges, which led him to serve another four years. He died in 2013 when he was 88 years old, having spent exactly 25% of his life in prison.[108]

RIGHT TO A JURY TRIAL

U.S. Constitution Article 3 § 2: The Trial of all criminals ... shall be by jury ....

Amendment 6: ...the accused shall enjoy the right to a speedy and public trial, by an impartial jury ....

Contrary to what some believe, the jury trial is not unique to the United States.[109] In fact, it has a long and storied history. The earliest known jury trials occurred in ancient Athens and Rome, where juries consisted of hundreds, sometimes thousands of citizens. They would hear the evidence and vote, with the decision going to the majority.

When the American colonies declared their independence from Britain and established their own government, they wrote into the Constitution the right to a jury trial, which was familiar to them as under English law, juries were a

---

[108] Estes was also famous for coming forward claiming to have key evidence tying former President Lyndon B. Johnson to the Kennedy Assassination and eight other murders, a position he maintained in a book he co-authored, which was published in France, and for which he was paid several hundred thousand dollars.

[109] We do use jury trials much more than most, if not all, other countries, however, because of their wide availability in civil cases.

regular feature.

The right to a jury trial is not absolute, however. As a general rule, the right is available to all criminal defendants, though there are exceptions:

- Criminal defendants not accused of "serious crimes," meaning the maximum imprisonment does not exceed six months;
- Criminal defendants who are juveniles;
- Members of the military (it varies); and
- Where the defendant waives the right to a jury trial (not available in all states).

Although juries are commonly understood to consist of twelve jurors, that is not universally true.[110] Federal courts use twelve-person juries, and most state courts require twelve for felony cases. The issue of how big a jury must be to provide constitutional protection to the accused has been considered a number of times. The U.S. Supreme Court has approved a six-person jury and rejected a five-person jury, and consequently, every jury in the U.S. for criminal cases consists of between six and twelve members.

Each state has created different requirements for unanimity in the verdict, which the U.S. Supreme Court has also had a few occasions to consider. Although states must no longer require juries to return unanimous verdicts, if the jury is smaller, like a six-person jury, the verdict must be unanimous. The Supreme Court concluded in one case that a state's 9-3 verdict threshold (75% majority vote) is constitutional, and in another case, that allowing an 8-4 guilty verdict (66 $\frac{2}{3}$% majority vote) to stand is not.

---

[110] It used to be that juries had to have twelve members.

RIGHT TO COUNSEL

Both the Fifth Amendment right to counsel[111] and the Sixth Amendment right to counsel[112] are constitutional trial rights enjoyed by criminal defendants. However, as they were already discussed at length above, we will not dive into them again here.

RIGHT TO CONFRONT WITNESSES

Amendment 6: In all criminal prosecutions, the accused shall enjoy the right to... be confronted with the witnesses against him.

The Confrontation Clause of the Sixth Amendment—the one that endows defendants with the right to confront their accusers—was not a right invented by the Founding Fathers. The concept of the accused confronting her accusers to defend herself is an ancient one and is explicitly referenced as Roman custom in the New Testament: "It is not the manner of the Romans to deliver any man to die, before that he which is accused have the accusers face to face and have licence to answer for himself concerning the crime laid against him."[113]

The Confrontation Clause ensures the defendant an opportunity to come face-to-face with the witnesses: the word confront, which finds its roots in the Latin *con* from *contra* meaning "against" or "opposed," and *frons*, which means

---

[111] *See* page 125.
[112] *See* page 128.
[113] Acts 25:16.

"forehead." Thus, the right of confrontation is quite literally the right to require the accuser's testimony to occur face-to-face with the accused.

Shakespeare understood this at the end of the sixteenth century when he wrote this line from *Richard II*:

"Then call them to our presence—face to face, and frowning brow to brow, ourselves will hear the accuser and the accused freely speak...."[114]

The right to a face-to-face confrontation comes with it the ability to cross-examine the witness to probe into their credibility. An artful cross-examination is not aimed merely at uncovering mistruths, but at undermine the efficacy of a witness' testimony by casting doubt on the witness' perception, memory, clarity, and sincerity.

Indeed, even the most truthful witness may be conveying false information. Faulty perception, which can have several causes, including external or internal impairment of the senses (everything from bad eyesight, poor lighting, sun in the eyes, excess noise, or even distraction) can challenge a witness's ability to accurately perceive the events that they are describing under oath.

Similarly, exploiting the imperfections of memory can be an effective method of discreditation. Most witnesses profess to remember the important details (important to the case) very clearly, but when pressed about other details, are either patently unsure or demonstrably overconfident.

Clarity has to do with a witness' ability to convey accurately what was observed. When questioned about their word choice—used in the comfortable confidence of direct examination—and pointedly asked to explain their testimonies,

---

[114] *Richard II*, Act 1, Sc. 1. *See also Coy v. Iowa*, 487 U.S. 1012, 1016, 108 S. Ct. 2798, 2801, 101 L. Ed. 2d 857 (1988).

witnesses will often either recant or discredit themselves with sheepish explanations.

The last impeachment stratagem is perhaps the most well known, though perhaps not by the name "sincerity." Sincerity is a measure of honesty, which can be disproven by revealing bias, prejudice, interest, or corruption.

Like other "guarantees," the Confrontation Clause has exceptions.

Forfeiture by Wrongdoing

Many a television show or movie shows a critical witness, enrolled in the witness-protection program, awaiting trial while cowering in a hotel room guarded by police. Such extremes became necessary to mitigate the danger that the defendant would be able to locate the witness and prevent them from testifying.

In the movies, the survival of the witness is critical to convict the defendant. In real life, it's not that simple. If the defendant were to murder the only witness in the case, for example, that does not guarantee the defendant's acquittal. In these cases, the testimony of the witness—recorded before they were killed—can be shown to the jury even though the defendant cannot confront the witness or cross-examine them.

This rule ensures serves a dual purpose: it ensures that critical testimony will be available even if the witness is not, and it discourages violence to witnesses by removing the reason for it.

Dying Declarations

The last words of a dying person are given considerable weight. The Latin maxim, *nemo moriturus praesumitur mentiri*, "no one on the point of death should be presumed to be lying" forms the rationale for this exception, which applies where the unavailable witnesses made a statement relating to the circumstances of their impending death while under the genuine belief that their death was imminent.

This is a popular media trope—the last act of the victim left some clue as to the identity of her murder.[115] This also happens in reality.

In February 2018, highway drivers in a rural section of Alameda County, California happened upon a bloody 19-year-old woman who was crawling toward Tesla Road. An air ambulance arrived to find her beaten and covered in stab wounds, and police were able to piece together that the attack happened 100 yards off the highway.

The victim, identified as Lizette Cuesta, used the last of her strength to name her neighbors Daniel Gross and Melissa Leonardo, whom the police immediately arrested.

Bucking his right to remain silent, Gross participated in a television interview with the local station where he admitted to the stabbings but claimed he was acting in self-defense when Lizette hit him in the ribs with her elbow, after which he just "snapped."

There were more than twenty stab wounds in Lizette's head, neck, and back.

Should this case go to trial, the jury will likely get to hear Lizette's dying declaration, as spoken through the person who heard her say them, which would normally be hearsay and contrary to the protections of the Confrontation Clause.

---

[115] In the media, it is never so simple as just naming the murderer clearly. That would defeat the purpose of having a brilliant detective solve the crime.

Victims of Abuse Who Are Children

Because of the special circumstances that attend children testifying against their abusers, and the chilling effect it would have on the vulnerable witnesses to require them to face the source of their psychological trauma, this situation forms another exception to the right of confrontation.

This is a limited exception. Defendants are still permitted to be privy to the child's testimony and cross-examine the witness, though not face to face. For example, the child may be in the courtroom with the judge and attorneys while the defendant is watching from closed-circuit television.

BURDEN OF PROOF

The United States, as many nations before it and since, recognizes and guarantees a presumption of innocence—that the accused is innocent until proven guilty.

Perhaps surprisingly, the Constitution does not guarantee, discuss, or even mention this right.[116] It was a right that pre-existed the birth of the United States and was built-in to the justice system with our adoption of English common law.

It was not until 1895 that the U.S. Supreme Court issued a decision explicitly reaffirming its place in American jurisprudence: "The principle that there is a presumption of innocence in favor of the accused is the undoubted law, axiomatic and elementary, and its enforcement lies at the foundation of the administration of our criminal law."[117]

The proof that the prosecution must bring is that which leaves no reasonable doubt in the minds of the jurors as to the

---

[116] Some have inferred this right from the Due Process Clause of the Fifth Amendment.
[117] *Coffin v. United States*, 156 U.S. 432, 453, 15 S. Ct. 394, 403, 39 L. Ed. 481 (1895).

guilt of the accused.

The term "reasonable doubt" is an important one. The prosecution is not required to erase all doubt or bring certainty to the minds of the jurors. But if the jurors have not resolved all reasonable doubts of the defendant's guilt—doubts supported by common sense—then the jury should acquit.

The defense team in the Casey Anthony case created a notable illustration of what reasonable doubt means.

Casey Anthony was the mother of a two-year-old named Caylee. Both resided in Florida with Casey's parents, George and Cindy. In June 2008, Casey left the home with Caylee, which was the last time the grandparents would see their granddaughter.

Over the next several weeks. Cindy called Casey several times and asked to see her granddaughter and was continually rebuffed. Casey made several excuses for why Cindy could not see Caylee—Casey was too busy with work, Caylee was with the nanny, Caylee was at a theme park, Caylee was at the beach. The name Casey attributed to her nanny—Zenaida Fernandez-Gonzalez, who was later questioned—had never met Casey, Caylee, or anyone else in Casey's family or anyone else who even knew Casey.

Twenty-seven days after Casey left, her parents received notice that Casey's car had been impounded. When George went to the tow yard to reclaim the vehicle, he smelled what he thought was a decomposing body in the trunk, though other than a bag full of trash, the trunk was empty. Mortified by this discovery, Cindy called the police and reported Caylee missing.

Police investigators interviewed Casey, who claimed that the nanny had kidnapped her daughter six days previously, but that she had been too afraid to call 9-1-1. Prosecutors were not convinced and charged Casey with first-degree murder, though she protested her innocence throughout the trial.

Six months later after Casey's arrest, in a grove near Casey's home, investigators found Caylee's decomposed corpse, wrapped in a blanket and placed in a trash bag. Caylee's body had duct tape near the front of the skull.

At trial, the prosecution presented several pieces of evidence to the jury, including the following:

- A piece of hair from the trunk of the car that, on a microscopic level, resembled Caylee's hair;
- Google searches on a computer accessible to Casey with search terms "neck breaking" and "how to make chloroform";
- "Zanny the Nanny" was a fabrication made up by Casey, as no such person existed;
- A poster on Casey's ex-boyfriend's MySpace page that said "Win her over with chloroform";
- Photographs of Casey at a nightclub with a tattoo she got three weeks after Caylee's death that said "Beautiful Life" in Italian; and
- A Winnie the Pooh blanket, known to be Caylee's and used at her grandparents' home, found with Caylee.

The defense presented an alternative theory: George found Caylee drowned in the pool, and to prevent neglect charges and to protect his daughter, he covered up the death. Casey did not report the crime and tried to hide her pain based on a habit of doing so from a lifetime of sexual abuse at the hands of her father, George, and brother, Lee.

The defense sought to discredit Casey's father, George, by alleging an affair with one of the search volunteers;

presenting evidence that George buried family pets in the same manner as Caylee was buried—wrapped in a blanket, placed in a bag, and wrapped with duct tape; suggesting that George had sexually abused Casey throughout her life; and revealing George's suicide note and suicide attempt, which occurred after his granddaughter went missing.

The defense emphasized that the police did not present any DNA, bodily fluids, or fingerprints belonging to Casey on anything found with Caylee's body. And if Caylee could outwit the cops by removing all traces of forensic evidence, it would not make sense that she would leave the body 19 feet from the road.

They highlighted the fact that Kronk—who had found the body—was not a witness the prosecution called, leading them to wonder why the prosecution would not want the jury to hear from him. The defense called him to testify, and there was some confusion about when he found Caylee's remains, which was cleared up when his son contradicted Kronk's testimony.

In closing arguments, the defense insisted, "this was an accident that snowballed out of control."

The defense also made an effective argument regarding the burden of proof: "The burden rests on the shoulders of my colleagues at the state attorney's office." The defense presented a "burden of proof" chart outlining a spectrum of doubt and the corresponding verdict:

GUILTY	GUILT BEYOND A REASONABLE DOUBT
NOT GUILTY	STRONG BELIEF
	GUILT LIKELY
	PROBABLY GUILTY
	POSSIBLY GUILTY
	SUSPECTED
	PERHAPS
	MAY NOT BE
	POSSIBLY NOT
	UNLIKELY
	PROBABLY NOT
	LESS THAN LIKELY
	HIGHLY UNLIKELY
	BELIEVED NOT GUILTY

With an understanding of the burden of proof, and reasonable doubt planted in their minds, the jury returned a not-guilty verdict for the murder of Caylee Anthony.

CHAPTER SIXTEEN

Sentencing and Punishment

THE HISTORY OF SENTENCING IS a long one with a slow yet steady evolution. The death penalty plays a prominent role in the development of sentencing systems. Almost 4,000 years ago, the Code of Hammurabi enumerated 25 different crimes that carried the penalty of death. Death was also a consequence of a number of crimes in colonial America (and still is to this day).

From the government's perspective, death was a simple solution to the age-old problem of crime. It was inexpensive, [118] final, and carried with it zero risks of criminal recidivism. And in theory, knowing that death followed those who chose to break the law had a deterrent effect on the rest of the population.

Two prominent Pennsylvanians—both Benjamins and

---

[118] It was inexpensive before due process rights. It is much more expensive now.

both signers of the Declaration of Independence—Dr. Benjamin Rush and Benjamin Franklin, challenged the notion that the death penalty serves as a general deterrent, arguing instead that the existence of the death penalty actually increased the rate of crime. As a result of their efforts, Pennsylvania not only was the first to create degrees of culpability for murder bust also led the way in repealing the death penalty for all crimes but first-degree murder. This occurred in 1794.

Sentencing someone to something other than death created the problem of not knowing what to do with the criminals. Before the early 19th century, penitentiaries were uncommon, and their emergence as a part of the fabric of social justice made imprisonment a viable alternative to death.

Another attribute of the sentencing system in colonial America that distinguishes it from most of today's courts, is that in colonial America, the jury both rendered a verdict and handed down the sentence. Today, in 41 states and the District of Columbia, sentencing is the exclusive province of the judge. In four states, the jury recommends a sentence, but the judge has the final say. Only in Arkansas, Missouri, Oklahoma, Texas, and Virginia do juries still sentence the criminal defendant.

SENTENCING RIGHTS

Amendment 8: Excessive bail shall not be required, nor excessive fines imposed, nor cruel and unusual punishments inflicted.

Since the earliest days of our country, punishments have evolved beyond just death or incarceration in a government facility. Punishment options are now wide and varied and may include the payment of a fine, payment of restitution to the victim, and probation.

The Constitution guarantees that whatever the punishment, it will not be "**cruel and unusual**." But because the Founding Fathers did not explain what makes a punishment cruel or unusual, the Supreme Court has had to interpret it and has defined the term using a test of **proportionality**—a measure of whether the punishment fits the crime.

Proportionality in sentencing requires an analysis of three factors: (1) the seriousness of the offence as compared to the severity of the punishment; (2) a comparison to other sentences imposed for similar crimes in the same jurisdiction; and (3) a comparison to sentences imposed for the same crime in other jurisdictions.

But even while staying within these constitutional strictures, judges have wide latitude when it comes to sentencing. If a jury renders a guilty verdict (or a judge finds the defendant guilty in non-jury cases), the judge[119] considers mitigating and aggravating factors before imposing the sentence.

The mitigating and aggravating factors are facts providing context to the crime. A **mitigating factor** is a fact that supports a more lenient sentence, like a sincere expression of contrition and regret, status as a victim of abuse as a child, or poor educational background. An **aggravating factor** is the opposite—a fact that supports the imposition of a harsher sentence, like prior criminal convictions or remorseless affect during trial. Aggravating factors must have been part of trial and proven beyond a reasonable doubt to be considered in making the sentence harsher.

The judge that sentences the defendant can choose what

---

[119] Or the jury in the minority of jurisdictions where the jury imposes the sentence.

weight, if any, to give to these sentencing factors. Federal Judge James Mahan, for example, presided over a case in the District of Nevada where lawyer Gerry Zobrist pled guilty to participating in a mortgage fraud conspiracy with such a broad scope that it contributed to Nevada's real estate crash in 2008.

Zobrist was a family man and popular in the community. Before the sentencing hearing, many of his family members and friends wrote letters to the judge on his behalf, pleading that the judge be lenient. At the sentencing hearing, the courtroom was filled with these supporters, hoping to show the sentencing judge that Zobrist had a strong support system.

Judge Mahan, weighing the gravity of the harm with the pleas of supporters, decided on a harsh sentence. Zobrist was sentenced to 87 months in prison, required to pay $31 million in restitution, and will have to subject himself to 5 years of supervision once he is released.

If the defendant pleads guilty or is convicted of multiple crimes, multiple sentences will be imposed, and the judge will get to choose whether the defendant will serve the sentences concurrently or consecutively. **Concurrent** sentences are served at the same time, and **consecutive** sentences are served one after another. A defendant who receives a five-year sentence, a three-year sentence, and a one-year sentence will serve a total of five years if they are concurrent, and nine years if they are consecutive.

In response to an apparent disparity in punishments, and in an effort to promote more uniform and fair punishments, Congress established the federal sentencing guidelines that more or less reduce sentencing to a mathematical formula. Judges use these guidelines to fit the conduct of the crime into one of 43 offense levels and compare that with the defendant's criminal history, which factors make up the rows and columns, respectively, of a sentencing table. The judge then sentences the defendant within the range set forth in the table. These guidelines were initially mandatory when established in the 1980s, but the Supreme Court relegated them to "advisory" status in 2005. Thus, judges may choose to, but do not have to, follow the sentencing guidelines.

A criminal defendant may be subjected to probation for all or part of the sentence. During criminal **probation**, the "probationer" is subject to certain restrictions and government supervision. A probation officer is assigned and ensures that the probationer makes restitution payments, stays away from drugs, does not commit any crimes, gets and keeps a job, and fulfills these or other conditions of the probation. The period of probation counts toward the total length of the sentence as long as the probationer meets the conditions of probation.

Parole is similar to probation. **Parole** occurs when prisoners are released early, serving the rest of their sentence outside the prison on terms similar to probation.

A defendant who is dissatisfied with the verdict or sentence can appeal to a higher court. A prosecutor cannot. Thus, while a conviction may just signal the beginning of a long appeals process, an acquittal (like with O.J. Simpson and Casey Anthony) is final.

Index

Index

A

abandoned property, 102
accessory, 28, 29
accessory after the fact, 28, 29
accessory before the fact, 28, 29
accomplice, 29, 30, 31, 32, 40
acquittal, 9
active possession, 19
actual cause, 24, 25
actus reus, 16, 17, 18, 20, 23, 24, 29, 34, 35, 38, 40, 45, 46, 67, 69, 77, 85
affidavit, 107, 108
Affordable Care Act, 4, 5, 6, 7
agency, 10
aggravated battery, 46, 59
aggravating factors, 15, 170
aiding escape, 29
armed robbery, 30
arraignment, 137, 146
arrest, 11, 13, 54, 83, 85, 86, 91, 104, 105, 106, 107, 108, 110, 111, 112, 113, 114, 115, 117, 136, 137, 139, 141, 142, 152
arson, 14, 22, 36, 85
asportation, 67

assault, 13, 21, 26, 28, 43, 45, 46, 47, 48, 50
attempt, 8, 34, 35, 36, 38, 40, 92
attenuation, 123, 124

B

bail, 38, 139, 140, 141
battery, 17, 20, 21, 26, 28, 36, 43, 45, 46, 47, 48, 50, 59
beyond a reasonable doubt, 8, 9, 16, 167, 170
bill, 3
binding over, 142
blackmail. *See* extortion
breaking and entering, 28, 81
bribery, 87, 91, 92
burglary, 13, 21, 28, 68, 81, 82, 83, 84, 85

C

capital offense, 2, 15
civil law, 8
complaint, 104, 139
concurrence, 23, 24
concurrent sentence, 171
confessions, 132, 133, 134

Congress, 2, 3, 8
consecutive sentence, 171
consent exception, 112, 117
conspiracy, 13, 21, 34, 39, 40, 41, 91, 92, 171
Constitution, 74, 96, 98, 115, 123, 128, 147, 153, 157, 163, 170
constructive possession, 19
criminal homicide, 51, 52
criminal law, ii, 1, 8, 11, 13, 95
cruel and unusual punishment, 170
curtilage, 99, 100

D

death penalty, 136, 168, 169
deterrence, 11, 12, 34, 169
discovery, 17, 106, 107, 124, 148, 152
Due Process, 134, 150, 163
Due Process Clause, 134, 163
dying declaration, 161, 162

E

education, 11, 13, 34, 134
elements, 16, 22, 27, 34, 45, 56, 72, 81
embezzlement, 21, 68, 69, 70, 73, 78
emergency exception, 113
evanescent evidence, 113, 119
exclusionary rule, 121, 122, 123, 125, 126
exculpatory evidence, 144, 149, 150
excusable homicide, 51
exigent circumstance, 113, 118
extortion, 73, 74, 75, 79

F

factual impossibility, 36
false imprisonment, 43, 45, 50, 56, 57
false pretenses, 21, 70, 72, 73, 77, 78
felony, 14, 15, 21, 29, 38, 51, 52, 67, 81, 83, 84, 113, 143, 150, 158
felony-murder rule, 51, 52, 150
Fifth Amendment, 21, 122, 128, 131, 133, 137, 143, 156, 159, 163
First Amendment, 154
first-degree murder, 52
fly-over exception, 100

forgery, 21, 28, 77, 78, 79
Fourteenth Amendment, 98, 128, 133, 134, 156
Fourth Amendment, 97, 98, 100, 101, 102, 103, 104, 105, 107, 110, 111, 112, 116, 117, 118, 121, 128
fraud, 60, 61, 62, 68, 71, 81, 154, 155, 157, 171
fraud in fact, 61
fraud in the inducement, 61
frisk, 118
fruit of the poisonous tree, 122

G

general intent, 21
good faith, 122, 123, 125
grand jury, 125, 132, 142, 143, 144, 145
grand larceny, 28, 67, 68

H

harboring a fugitive, 29
homicide. *See* murder
hot pursuit, 112, 118

I

impossibility, 36, 38
incapacitation, 11, 34
inchoate, 26, 34, 38, 39, 92
independent source, 123
indictment, 137, 142, 143, 144, 145, 146, 147
inevitability, 123
information, 137, 144
initial appearance, 139, 141, 142
intent, 20, 21, 22, 23, 29, 31, 34, 35, 36, 38, 39, 45, 47, 51, 58, 67, 69, 72, 77, 79, 81, 83, 85, 89
interrogation, 127, 128, 129, 130, 132, 133, 134, 137
intoxication manslaughter. *See* vehicular manslaughter
Investigative grand jury, 144
investigatory stop, 104, 105, 115
involuntary manslaughter, 51, 53

175

J

justifiable homicide, 51

K

kidnapping, 17, 20, 21, 43, 45, 50, 57, 58, 129, 131
knowingly, 23

L

larceny, 21, 67, 68, 69, 70, 71, 72, 73, 78, 79
larceny by trick, 67, 71, 73
legal impossibility, 36

M

malice, 22, 23
malice aforethought, 51
malum in se, 11, 14, 53
malum prohibidum, 11, 14
mayhem, 43, 45, 48, 49, 50
mens rea, 16, 20, 22, 23, 24, 29, 34, 38, 39, 45, 46, 63, 67, 69, 77, 85
merger, 40
Miranda Rights, 131, 132
misdemeanor, 14, 15, 53, 67, 83, 84, 113
mitigating factor, 15, 170
motor vehicle exception, 112, 115
murder, 1, 9, 11, 14, 17, 20, 22, 24, 25, 26, 28, 30, 33, 35, 36, 38, 39, 43, 45, 51, 52, 53, 58, 150, 161, 162, 165, 167, 169

N

negligence, 23, 45, 53
neutrality, 106
no bill, 144
no contest, 146
nolo contendere. *See* no contest

O

Obamacare. *See* Affordable Care Act
obstruction of justice, 29

omission, 17, 18
open fields, 99, 100
overt act, 34, 35, 39, 40, 41

P

parole, 172
particularity, 108
PATRIOT Act, 4, 5, 6, 7
perjury, 87, 88, 89, 90
personal recognizance, 140, 142
petit larceny, 67
Pinkerton liability, 41, 52
plain view exception, 112, 117
plea bargain, 147, 152
plea negotiations, 137
possession, 18, 19, 26, 31, 69, 70, 71, 76, 104, 115
preliminary hearing, 137, 142, 143, 144
preponderance of the evidence, 8, 9
presumption of innocence, 88, 163
principal, 27, 28, 29, 30, 102, 154
principals in the first degree, 28
principals in the second degree, 28
probable cause, 103, 106, 107, 108, 112, 113, 115, 117, 118, 141, 142, 143, 144
probation, 11, 54, 63, 170, 172
procedural law, 13
proportionality, 170
prosecutor, 8, 16, 21, 58, 113, 122, 123, 138, 139, 143, 144, 147, 148, 149, 150, 172
proximate cause, 24, 25
purposely, 23

R

rape, 14, 18, 31, 35, 59, 60, 61, 62, 64, 65, 129, 130, 131
reasonable doubt, 163, 164, 167
reasonable expectation of privacy, 99, 106, 108
receipt of stolen property, 75, 76, 79
recklessly, 23
rehabilitation, 11, 13, 34
retribution, 11, 12, 13
right against self-incrimination, 9
right to a jury trial, 146, 153, 157, 158

right to a public trial, 153, 156
right to a speedy trial, 147, 148
right to an attorney, 132, 133, 135, 136, 137, 139, 144, 153, 159
right to confront witnesses, 146, 153, 159, 161, 162
right to remain silent, 131, 132, 133, 146, 153, 162
robbery, 13, 30, 35, 37, 38, 72, 73, 79, 115, 129, 146, 149

S

search, 97, 98, 99, 100, 101, 102, 103, 104, 105, 106, 107, 108, 110, 111, 112, 114, 115, 117, 118, 119, 120, 121, 122, 123, 125
search incident to an arrest, 114
second-degree murder, 51
seizure, 57, 98, 99, 104, 105, 106, 108, 111, 112, 117, 118, 122
sentencing, 2, 15, 26, 54, 63, 146, 150, 168, 169, 170, 171
sentencing guidelines, 171
sex offenses, 59
Sixth Amendment, 122, 131, 135, 136, 137, 156, 159
solicitation, 34, 38, 39, 40, 91
specific intent, 21
statutory rape, 22, 63
stop, 118
stop and frisk, 112, 117
strict liability, 22, 23
subornation of perjury, 87
substantive law, 13, 14

summons, 139
surreptitious remaining, 83

T

torture, 44, 47, 49, 133, 134
transferred intent, 46
true bill, 144

U

USA PATRIOT Act. *See* PATRIOT Act
uttering a false instrument, 77, 78

V

vehicular manslaughter, 53
veto, 4
vicarious liability, 20
voluntary confession, 134, 135
voluntary manslaughter, 51, 52, 53

W

warrant, 54, 99, 100, 101, 102, 103, 104, 105, 106, 107, 108, 109, 110, 111, 112, 113, 114, 115, 116, 117, 118, 119, 123, 125, 139, 142
warrant exception, 99, 101, 102, 105, 106, 110, 114, 119
wire fraud, 28
wiretapping, 108, 109
withdrawal, 31, 32, 41

Made in the USA
Las Vegas, NV
03 November 2025

33574430R00105